SCIENCE DISCOVERY ACTIVITIES KIT

Ready-to-Use Lessons & Worksheets for Grades 3–8

Frances Bartlett Barhydt

THE CENTER FOR APPLIED
RESEARCH IN EDUCATION
West Nyack, New York 10995

Library of Congress Cataloging-in-Publication Data

Barhydt, Frances Bartlett.
  Science discovery activities kit.

  1. Science—Study and teaching (Elementary)
2. Activity programs and in education.   3. Creative
activities and seat work.   I. Title.
LB1585.B28   1989   372.3′5044   89-712
ISBN 0-87628-785-2

ISBN 0-87628-785-2

THE CENTER FOR APPLIED
RESEARCH IN EDUCATION
BUSINESS & PROFESSIONAL DIVISION
A division of Simon & Schuster
West Nyack, New York 10995

For my lifetime teachers:

My parents, Frances and James Bartlett, who taught me the values of learning.

My methods professors, Bruce Joyce and Elizabeth Crook, who taught me how to teach.

My husband, James Barhydt, who taught me that I could write.

I Hear . . . and I Forget
I See . . . and I Remember
I Do . . . and I Understand

Ancient Chinese Proverb

About the Author

Frances Bartlett Barhydt earned a B.S. degree in education at the University of Delaware and an M.A. in teaching at Central Michigan University. Her teaching experience is varied. It is unique in that it includes experience at the elementary, junior high, and university levels, along with on-camera teaching in the Walt Disney Educational Media Company film *Chemistry Matters*.

In both 1982 and 1984, Mrs. Barhydt was recognized by the National Science Teacher's Association (NSTA) as "representing excellence in the area of elementary science for the state of Delaware." In 1985, she was awarded a certificate of recognition for "outstanding contribution in the teaching of elementary science" by the Council for Elementary Science International of NSTA.

Mrs. Barhydt has also written a number of filmstrip and cassette publications, including "Fort Delaware: Yesterday and Today," as well as articles on science education for the *DSEA Journal, Instructor,* and *Science and Children.*

About This Book

For 25 years I have been having a love affair with science education. Introducing children and adults to the essence of science is rewarding and at times thrilling. To see the excitement of discovery on a child's face or the loss of fear on that of an adult are wonderful experiences that I hope you will encounter.

I am not a scientist. I am an elementary educator who has learned the educational dividends that can be obtained by introducing the investigative nature of science to elementary, junior high, and university science methods students. Safe, scientific, investigations provide students with a variety of experiences with discovery, problem-solving, and logical and creative thinking. These investigations gradually allow students to be active participants and to become responsible for their own learning. They present many opportunities for students to be successful and to develop positive attitudes and healthy self-concepts.

The guided discovery approach with hands-on activities has been the most rewarding approach that I have tried. It is rewarding because it effectively meets the needs of students with a wide range of academic abilities and social development. In other words, it is not exclusively for the brightest and best behaved. It has helped me motivate and challenge students in:

- grades three through seven
- schools in small towns, suburban areas, and in inner-city classes labeled "discipline problems"
- classes composed of students with varying abilities, including those identified as academically gifted and talented, socially and emotionally maladjusted, and learning disabled
- classes composed of students from different socioeconomic, ethnic, and racial backgrounds

The guided discovery approach requires time and planning to develop the three major aspects of the lesson: the inquiry, the activities to support the inquiry, and a management system that provides a blueprint for safe investigations and acceptable behavior.

My first class taught me the value of well managed, guided discovery lessons. It became my class after I graduated in January, the original teacher having resigned in December. The principal who hired me explained that the boys caused the largest plumbing bill in the history of the school by flushing marbles down the toilets. Boys and girls threw crayons in the hanging lights and watched to determine how long it would take them to melt.

Determining the melting time of crayons could be an interesting investigation and could lead to practical applications such as candle making. However, these students were lacking respect for the school, self-discipline, and safety standards.

They had demonstrated that they were curious and preferred to manipulate objects. Thus, my main goal was to provide highly interesting but teacher-controlled investigations that would motivate them to participate and follow safety regulations. They loved investigating, responded to guided discovery, and welcomed an environment in which they could experiment within limits. As their self-discipline improved, the investigations gradually became less teacher-controlled and limits were determined by safety, logic, materials that we were able to obtain, and places that we were able to visit. Investigations gradually evolved into ones that were more student-centered and open-ended.

To teach guided discovery science lessons with hands-on activities in the elementary and middle school, you do not have to be a genius or a science major. You need to:

- understand the kinds of experiences that are best suited for students at different levels of their development
- understand basic human needs of safety, belonging, and self-esteem
- understand the inquiring nature of science
- be willing to be a discovery learner
- build a repertoire of investigations with which you feel comfortable
- be an effective questioner
- be an effective manager

Many of the teachers, student teachers, and science methods students who observe in my elementary science classes are interested in initiating or expanding the inquiry process in their own classes. The questions that I am asked most frequently are:

"How do you get started with hands-on activities?"

"How do you keep control?"

"What do you do with disciplinary problem students?"

"How do you know what to teach?"

The purposes of the *Science Discovery Activities Kit* are to:

- answer these questions
- build your knowledge of general science
- provide examples of discovery lessons
- provide guidelines for a science management system
- provide guidelines for a hierarchy of discipline
- help you initiate and develop the inquiry process in your classroom

Elementary science education is no longer nature study or assigning readings and answering the questions at the end of the chapter. It is teaching students skills for learning how to learn, how to question, how to investigate, and sometimes how to recognize that there isn't enough evidence to come to a valid conclusion.

Science isn't exclusively for the scientist. The problem-solving skills of science are valuable, lifetime tools that are important for everyone.

Fran Barhydt

How to Use This Resource

To initiate the guided discovery approach with hands-on activities:

1. Read Sections 1 and 2.
2. Skim the rest of the *Kit*.
3. Identify a section that contains subject matter with which you feel comfortable.
4. Read that section carefully to select the ideas and activities that best meet the needs of your class and match the supplies that you can obtain.
5. Duplicate appropriate information sheets and/or worksheets, but use them as class or group learning experiences, not reading exercises.
6. Keep your initial discovery lessons with hands-on activities simple and teacher-controlled.

To develop and enhance the inquiry process in your class:

1. Teach a few teacher-controlled, guided discovery lessons while training your students to follow safety regulations and class procedures.
2. Develop a series of lessons around a theme that builds concepts, skills, and attitudes.
3. Develop a holistic, interdisciplinary unit of study which builds concepts, process skills, and attitudes.
4. Present a variety of investigations that gradually become student-centered and open-ended.
5. Teach students how to identify and work on their own areas of investigation.

Note: The first two lessons in each section appear to be simple. They are designed to provide every student with basic information on which to build and succeed. Each successive lesson builds content while using scientific process skills.

Contents

SECTION 4: INTRODUCTION TO CHEMICAL AND PHYSICAL CHANGES   79

SECTION 5: GRAPHS: GREAT GRIDS FOR COMMUNICATING   137

SECTION 6: THE HUMAN BODY: A CUT-AND-PASTE MODEL   165

SECTION 8: GUIDED DISCOVERY WITH ELECTRICITY   **257**

Acknowledgments

Many friends and colleagues have contributed their expertise and recommendations for the manuscript. Their contributions have strengthened this book. I am grateful to:

Mr. James Barhydt, editor and writer

Dr. H. Bruce Carrick, chiropractor

Dr. D. Robert Coulson, chemist

Mrs. Hilary H. Coulson, teacher and supervisor

Dr. James Culley, marketing researcher

Mr. Harry Hammond, history teacher and administrator

Mrs. Melody Hammond, special education teacher

Ms. Ann Leuthner, education editor

Dr. Paul Morgan, chemist

Mrs. Ellen Levine, physical therapist

Dr. Elizabeth A. Wier, elementary science methods professor

The Guided Discovery
Approach to Teaching Science

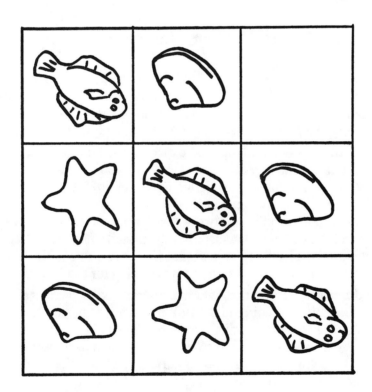

PART A
About the Discovery Approach

What Is the Guided Discovery Approach to Teaching Science?

The guided discovery approach is considered by most science educators to be the most effective way to teach science. It involves students in investigations with exploratory activities that lead them to draw valid conclusions, acquire skills, and understand concepts. The exploratory activities provide concrete experiences to help students understand and remember abstract ideas without rote memorization. These activities can include hands-on science labs, demonstrations, art projects, field trips, and readings. Students gather and record data, base conclusions on the data, and report their findings. The discoveries that students make are new to the students, not to the scientific community.

Why Teach the Guided Discovery Approach?

It works! It is effective with both students who can learn in a traditional approach and most of those who cannot. It provides students with science activities that help them learn basic science and academic skills. The experiences help many build vocabulary, improve reading comprehension, and improve measurement skills.[1]

It's exciting. When students make discoveries they get excited and frequently want to show one another and the teacher. With young students it's the "Wow" or "Hey" effect; students make discoveries and excitedly announce, "Wow! Look at that!", Or "Hey, Mrs. B. come look at this!"

With older students it is usually the "Oh" effect. It takes place when a concept is actually understood and a student says in a surprised voice, "Oh, that's what is meant by series and parallel circuits. I've had that before but this is the first time that I've understood it." These effects make learning exciting for the student and the teacher.

It is educationally sound. Questioning to make students think dates back to Socrates. Learning by doing and thinking about what you are doing have been urged by John Dewey and studied by Jean Piaget. Discovery learning has been studied and recommended by cognitive psychologists Jerome Bruner and Robert Gagné. Discovery learning provides experiences that help students understand, apply, and recall what they have learned.

---

[1] *What Research Says to the Science Teacher.* Mary Budd Rowe, ed. Washington, D.C.: National Science Teachers Association, Vol. 1 (1978) and Vol. 2 (1979).

It is practical. It provides the classroom teacher with an effective method to achieve required objectives and manage materials, activities, teaching/learning time, and discipline.

Its advantages are many and varied. It provides students with opportunities to move from manipulating concrete objects to thinking about and communicating abstract ideas. It provides students with a means for experiencing simple ideas, and moves them step-by-step to more complex concepts. As students acquire concepts and skills, they experience success with problem solving. Success encourages interest, willingness to try something new, and positive attitudes about science, school, and self. Accordingly, many of us who teach the discovery method find that it eases many learning and behavioral problems. Yet the discovery approach to teaching science is not the most widely accepted practice in our classrooms. Why?

Major Deterrents

The major deterrents to widespread practice of discovery teaching are teachers' concerns about discipline, equipment, covering required curriculum, lack of personal scientific knowledge, and disastrous experiences with hands-on activities. It is important to think about how these and other deterrents have influenced your philosophy of education because your philosophy determines what you choose to teach and how you teach it.

Other Limiting Factors

Another factor which limits the acceptance of discovery teaching stems from our own learning experiences. Consider that we generally teach the way that we were taught; most of us were taught abstract ideas with paper, pencil, and books in an academic setting.

When we are introduced to the inquiry method of teaching and discovery activities, it is usually through lectures and readings rather than actual experiences. Thus, it is easy to develop the misconceptions that lecturing and assigning readings are the most effective ways to communicate information.

Also limiting is the failure among teachers and administrators to continually ask, "What does research say to us as educators?" Decisions about curriculum and teaching should be, but rarely are, based on research. Another problem comes from a misunderstanding of what science is. Science is not a body of facts and laws for the intellectual elite. Science is the process of *inquiry*. Scientists utilize skills and concepts to logically and creatively investigate a problem or an idea in an organized and disciplined manner. They identify an investigation, design tests and controls, gather and record data, make conclusions based on the data and communicate those findings. Findings are communicated in oral and/or written presentations using a variety of visuals including charts, diagrams, and graphs. There is no *one* scientific method to be applied to all investigations and presentations. When science is taught as a process of inquiry, the student becomes the investigator gaining understanding of what science is about and what scientists do.

Scientists are not always right and are not perfect. They learn by trial and error as well as by reading, listening, and writing. They work hard to investigate problems and arrive at valid conclusions. Teachers and students do not always have to be right. They need to be willing to employ a variety of techniques to learn, and to work diligently to arrive at valid conclusions.

Unclear Goals

At present, there is an emphasis in the media and in educational literature to improve science and math teaching. Unfortunately, many interpret this to mean that there should be more science and math taught, and at an earlier age. More is not necessarily better. Teaching complex concepts to students before they are developmentally ready to understand those concepts leads to unproductive frustrations and misconceptions.

Goals for Science Education

The major goals of science education in the elementary and middle schools should be to:

1. Provide students with learning experiences that are commensurate with their cognitive and social development.
2. Teach lifetime skills for identifying and applying information to make valid conclusions and informed decisions.
3. Provide experiences that demonstrate how to transfer scientific, problem-solving skills to other academic and nonacademic areas.
4. Provide successful encounters with science that help students discover that they want to and are able to learn, as well as the mechanics of how to learn.
5. Provide opportunities to consider career choices.
6. Provide a quality foundation of skills and concepts enabling students to become scientifically literate.
7. Provide a part of the school week that is intellectually challenging and satisfying.

How to Meet the Goals for Science Education Through Discovery

To meet the goals of science education through discovery teaching, you need to do what all competent teachers do: plan, set short-term goals, organize, manage, and use a variety of teaching techniques to meet students' needs. But in the discovery approach, planning, organizing, and managing center around what is needed to help students investigate a problem or topic and reach valid conclusions.

How Do You Teach the Guided Discovery Approach to Science?

There are different forms of discovery teaching. Each form teaches investigations, but each varies in degree of teacher involvement, control, and management. It is important to have a clear understanding of the differences between two extreme forms of discovery teaching: guided discovery and pure discovery.

In the guided discovery approach the teacher:

- teaches student expectations and responsibilities for individual and group work
- identifies investigations by asking questions
- provides equipment and directions for using it to conduct the investigation or to answer the question
- provides guidelines and safety rules for classroom procedures
- is a resource person who provides guidance, information, and questions that help students solve problems and make discoveries
- directs students to work individually or in groups, on the same problem or on different problems
- monitors student work and behavior
- manages the learning experiences to ensure that the required curriculum objectives are met, valid conclusions are reached, and a safe learning environment is maintained

Teacher-controlled, guided discovery lessons are recommended for initial discovery lessons at the beginning of the school year. As students demonstrate that they can work in investigatory groups, you may move toward more student-centered and open-ended investigations, but they must remain teacher managed. As the students become more adept at working in lab and group activities, it will appear that you are managing less and less. Actually the students are becoming more responsible and are using the management guidelines that you have established and taught.

Pure Discovery

In the pure discovery approach, the teacher:

- does not always identify a specific investigation
- provides students with directions for safety reasons
- provides equipment

- directs students to investigate
- answers questions about safety regulations or directions, but does not answer information-seeking questions

In the pure discovery approach, the teacher may have difficulty meeting required curriculum objectives and student misconceptions may result because valid conclusions are not reached. Without guidance, students cannot be responsible for meeting curriculum objectives.

Pure discovery also requires a management system. Many have the misconception that pure discovery means "turn students loose" with equipment to see what they can discover. Think about what that would mean in your class. It invites chaos.

Pure discovery needs to be thought of as a goal possible for some students, not a starting point for an entire class. It is recommended for students who are self-disciplined, self-motivated, possess knowledge in scientific investigation skills, and have had many experiences with guided discovery learning. If you have a group of students who responds well to guided discovery teaching, you may want to try some pure discovery lessons. Only you can make that decision.

Consider the Two Discovery Approaches When Introducing Open and Closed Electrical Circuits

In the pure discovery approach the teacher:

- Reminds students to follow safety rules
- Gives each group one battery, one bulb, and one wire
- Monitors the students' behavior
- Answers no questions for information; except to answer a question with a question
- Collects the materials
- Conducts a discussion to ask the students what was learned

Caution: When specific investigations are not identified by the teacher, some will be identified by the students. Those identified by students do not necessarily meet your objectives, can be unproductive, and can lead to unacceptable behavior. Some students may choose to investigate:

1. How high will a battery bounce?
2. What is inside a battery?

This approach causes a typical problem: the students did just what they were told to do . . . investigate. This frequently causes problems and makes it difficult for the student to develop trust in the teacher. Don't expect students to discover your objectives or what you want them to investigate. Give clear directions. Identify the investigation and provide the knowledge they need to solve it.

In the guided discovery approach, the teacher:

- Identifies the investigation: Given one light bulb, one wire, and one battery, find at least four different ways to complete the system to make the light glow. As you discover a way to make the light glow, draw a model of the circuit. When you have drawings of four different circuits that make the light glow, tape them on the wall. If you finish early, see if you can find other ways to make the bulb glow.
- Assigns individual and group responsibilities.
- Directs selected students to distribute trays of supplies.
- Is a quiet resource person who sits in on each group. May ask, "Have you tried . . . ?"
- Monitors the class.
- Uses own system of management and discipline.
- Directs students to collect trays of supplies.
- Directs each group to explain one of its drawings.
- Introduces the terms "open" and "closed" circuits based on the lab experience and the drawings. In an "open" circuit, the light does not glow. In a "closed" circuit, the light glows.

Discovery Teaching Makes Effective Use of Traditional Techniques

Discovery and traditional teaching methods use the same teaching techniques of demonstrating, questioning, reading, lecturing, and discussing. To be a discovery teacher, it is helpful to understand the similarities and differences between the two methods. They differ in emphasis, order of activities, and role of the teacher.

In the traditional teaching method:

- The emphasis is on content, which is introduced through reading and mini-lectures.
- Demonstrations, discussion, questioning, and activities follow and support the introduction.
- The role of the teacher is that of an authority or giver of information; a "teller."

In the guided discovery approach to teaching:

- The emphasis is on the application of content and skills in investigations to arrive at valid conclusions.
- Content and skills are introduced by posing a problem and providing an exploratory activity to investigate the problem.

- Reading, minilectures, and discussions follow and support the initial investigation.
- The role of the teacher is that of a resource person and effective manager.

Discovery Learning

To develop and enhance your discovery teaching skills it is important to analyze a discovery learning experience. Remember, the discovery teacher:

1. Identifies a problem.
2. Provides an exploratory thinking experience.
3. Leads a discussion of what was learned.
4. Follows up with practice and reading.

Here Is a Discovery Experience for You

Try this discovery experience, analyze it, and consider using it for an introductory lesson. Your problem is to solve the Latin Square, Figure 1-1.[2]

1. Observe the Latin Square.
2. Decide what should go in the empty square.
3. Sketch your answer in the empty square.
4. In a phrase explain how you solved the problem.
5. Check your answer by looking for a different way to solve the problem. Do you get the same answer no matter how you solve the problem?

If we were in a discovery science class, we would discuss a variety of ways of solving the Latin Square and how each solution adds confidence to the validity of our answer. Our "discussion" would have to consist of comparing answers and explanations. To check your answer and compare explanations, see page 20.

Think About the Discovery Learning Experience

1. How did you feel when you realized that you knew how to solve this problem? (Most people enjoy success.)

2. Can students in grades three through eight solve this type of problem? (Absolutely.)

---

[2]David Lewis and James Green. *Thinking Better.* New York: Rawson, Wade Publishers, Inc. 1982.

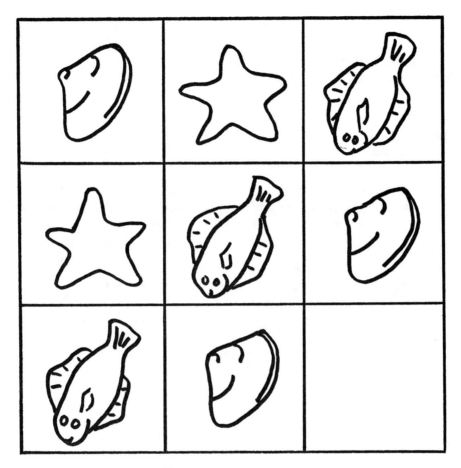

Figure 1-1. A Latin square

3. Does success depend upon traditional reading? (No.) Consequently, those who have difficulty reading on grade level have a good chance for success.

4. If a slow learner and/or a discipline problem student solves this problem, how will each of them react? (Students with learning and behavior problems enjoy success as much as anyone. However, they are usually not used to much success in school and may be surprised and minimize their achievements. They need continuous reinforcement just as other students do.)

To Apply This to Your Classroom

1. Draw a Latin Square on the board before students arrive. Copy the one in this section or make one of your own using any type of appropriate drawings such as flowers, trees or geometric shapes. Cover it with a pull-down map or screen until you are ready to use it.

2. Give specific directions such as, "We are going to do some thinking. It will require quiet. I'm going to uncover a problem on the board. When you have an answer and can explain it, raise your hand. Please do not call out because that interferes with your classmates' thinking."

3. Reveal the problem. Quietly wait for many students to signal that they have an idea. Call on one person to give an answer and an explanation. Demonstrate why that response is correct or incorrect. After the correct answer is discussed, ask for another method for arriving at the correct answer. Continue with a variety of explanations.

4. Direct students to use paper, pencil, and a ruler to make a Latin Square of their own to be due at the end of the class. Invite those who are interested in making other brain teasers for extra credit to consult some of the puzzle books in the class or in the school library. (Be specific in how you will count the extra credit.)

Is this really a discovery science lesson? Absolutely. This lesson doesn't use expensive scientific equipment. It isn't a hands-on experience, but it is a problem-solving experience. It teaches the quintessence of science: observation, searching for patterns, developing a tentative answer or inference, testing that inference, basing the conclusion on the data, reproducing the conclusion, and building confidence in the conclusion. Latin Squares and other brain teasers can be excellent foundation lessons for discovery science.

PART B
How to Succeed with the Guided Discovery Approach

Your success in implementing and expanding the guided discovery approach to teaching science depends upon:

• *Developing discovery teaching techniques* that reflect a respect for questioning and thinking. To do so:

1. Question rather than lecture.
2. Provide thinking time.
3. Refrain from judgmental statements during the discovery part of the lesson.

• *Developing a management system* that provides a safe environment in which discovery learning can flourish. To do so:

1. Teach school expectations and requirements.
2. Teach science and safety expectations and requirements.
3. Develop your own hierarchy of discipline.
4. Set up a file to provide non-lab, science lesson plans for substitute teachers.

The remainder of this chapter details how to develop discovery teaching techniques and a management system for those teachers who are new to the guided discovery approach.

How to Develop Discovery Teaching Techniques

Discovery teaching techniques depend upon developing the art of questioning. An atmosphere in which questioning and thinking are valued is created by the way you ask questions and the manner in which you respond to student answers and questions. There are three keys to demonstrate that you value questioning and thinking.

1. Practice Questioning Rather than Telling or Lecturing

This furthers the lesson, stimulates thinking, and identifies opinions. Ask open-ended or thought-provoking questions, such as:

• How can we determine which of these items will sink or float?
• Have you ever thanked a green plant for its life-giving products?
• Do you think there are any helpful bacteria?

2. Provide Students with Time to Think After You Ask a Question

This sounds deceptively simple but teachers ask an average of two or three questions a minute giving students only 9/10 of a second before making a remark or rephrasing the question.[3]

3. Do Not Make Judgmental Statements During the Investigative, Thinking Part of the Lesson

Students will become more interested in your reactions of reward, shock, or attention rather than thinking. Even positive statements meant to reinforce positive behavior interfere with thinking. Save such statements for the non-discovery parts of the school day.

Applying These Discovery Techniques

To think about applying these discovery techniques, consider the following introductory discovery lesson on magnetism. Set up a demonstration area with a magnet and four metal nuts and bolts, each with a label identifying its makeup: iron, steel, aluminum, and stainless steel. It is important to tell the students what the metal of each object is because they have no means of discovering its metallic contents. Ask, "How can we tell if any of these objects are magnetic? Raise your hand when you have an idea that we could try."

Then, give students time to think, call on someone who has his or her hand raised, and listen to that person's response, "They're all magnetic because they're all metal."

This is a common misconception, but do not correct it immediately.

This is also an example of a common nonscientific behavior among beginning science students. That is, jumping to conclusions rather than gathering data.

Respond, "That is your inference or what you think. How could you test your inference to find out if it is true or not?"

Note: The test is to determine whether or not the *inference* is true, not to see if the student is right or wrong.

A possible student response is, "Touch each one with a magnet and see if it will stick."

You then ask, "Would you like to try it?" If the student would, he or she is invited to perform the test. If not, ask for a volunteer.

To access student thinking and involve all students, direct everyone to make a prediction *before* the actual demonstration. Say, "If you predict that all of these items are magnetic, make two fists." Wait three seconds. "If you think they are not

---

[3] Mary Budd Rowe, *Teaching Science as Continuous Inquiry*. New York: McGraw-Hill, 1973, p. 273.

magnetic, open both hands." Wait. "If you think some are magnetic and some are not, make one fist and open one hand."

As each item is tested, the result is recorded on a chart on the board. See Figure 1-2. The test results reveal that these iron and steel items are attracted to the magnet and aluminum and stainless steel items are not. In addition to teaching about magnetism, the teacher is demonstrating that he or she listens to students, realizes inferences do not have to be perfect to be useful, bases conclusions on tests results, and values student participation and thinking.

**Identifying and Classifying Metals
That Are Magnetic and Not Magnetic**

Magnetic	Not Magnetic

Figure 1-2. Board chart

How to Develop a Management System for Guided Discovery Science

An effective management system for guided discovery science is one that provides the teacher with a frame of reference for working with all students and provides the students with clear expectations and a safe, orderly environment in which to learn.

Effective management makes it possible to eliminate some discipline problems before they arise, and to deal with others as they develop. No *one* system of management is best for all science classes. Consider the following recommendations for the eight steps to an effective management system and modify them to meet your needs.

1. Teach Traditional Expectations and Regulations

Students need to be taught what is expected of them. There are expectations set by school policy and those set by the classroom teacher. It is important to prioritize them and to teach the most immediate requirements, procedures, and routines during the first week of school. The sooner students understand and follow daily routines, the sooner they feel comfortable in their new environment and become responsible for their own behavior. Certain times of the school year require more attention to management than others:

- **The first week and especially the first day of school.** Plan to invest at least one half of your time teaching school and class rules, routines, requirements, and schedules. Distributing lists of rules and delivering a monologue on procedures is just as ineffective with teaching expectations as it is teaching academic concepts. Students need to be involved, solve problems, and apply information.
- **The first month of school.** After routines have been taught, you need to reinforce them with gentle reminders, thank-you's, and consequences when necessary.
- **Any time that you introduce a new type of science class,** such as a demonstration or a lab.
- **A few days before a school vacation.** See "Christmas," pages 80 and 110–118.
- **The last week and last day of school.** See "The last day of school," page 131.

2. Teach Science Class Expectations, Regulations, and Procedures

(A) Teach Science Expectations. In all science classes, no matter the grade level, student and teacher safety must be a priority. Each science classroom needs general and specific safety rules. The general rules need to be posted, taught the first day of school, and consistently followed throughout the school year. Examples:

Science Safety Rules

(1) Use supplies and equipment only after teacher gives directions and permission.
(2) Walk in the science room. Running can cause accidents.
(3) If you break something made of glass, *do not* clean it up. Immediately notify the teacher. The teacher will give you a bag and a marker. Write "broken glass" on the bag. The teacher will gather the glass, put it in the bag and tape it to the outside of the wastebasket.
(4) If you spill or make a mess, stop what you are doing. Clean it up. Then resume your investigation.

Specific safety rules and regulations, such as demonstrating the use of a piece of equipment, or procedures for using traditional supplies in the science classroom, need to be taught when they apply to a lesson for the first time. Examples:

Specific Safety Rules and Procedures

(1) To use an eye dropper, you need to hold it perpendicular to the desk.
(2) Scissors are to be stored with the shears closed and pointed downward in holders.
(3) When you finish using a ruler, place it flat on the desk in front of you.

(B) Teach Expectations and Procedures for a Science Demonstration.
For demonstrations to be effective, students must be able to see and to participate.
Teach students that you will:

- Give directions for seating arrangements before and after the demonstration.
- Involve voluntary lab assistants.
- Include everyone in making predictions before actual testing by using an "everybody" response; such as, "If you predict that this object is magnetic, show thumbs up." Wait three seconds for responses, then say, "If you predict that this is not magnetic, show thumbs down."

3. Teach Students to Work in Cooperative Groups

Teach students to help each other during group, practice, project, and study time. Assign students to groups of four. When the activity requires reading, assign the students by reading ability. Match a poor reader with a good reader. Thus, a typical group will be made of one excellent reader, one poor reader, and two average readers. Teach them to:

- work with the people in their group
- use an "indoor" or "library" voice
- be prepared to discuss what was learned

4. Consider Reinforcement and Competition

Sometimes students respond to group competition. There should be no competition within a group. Therefore, statements for positive reinforcement are directed at a group, not an individual.

- "Table 1 did the best job of cleanup today. Congratulations."
- With young students, "Table 1 did the best job of cleanup today, so they may line up first for lunch."
- "The group with the best improved test scores is . . ."

5. Teach Students to Work in Lab Groups

Teach students how to work in lab groups. Begin by assigning students to groups of four based upon how well they work together. If a student has a reading or handwriting problem, team that person with someone who is willing and able to help. Thus a typical lab group is composed of four students who work well together.

Explain that there are individual and group responsibilities. Each individual is expected to:

- Complete his or her own lab paper.
- Cooperate with group members.

- Help clean up the group area.
- Perform one of the group responsibilities.
- Stop whatever he or she is doing and give the teacher his or her attention upon a prearranged signal such as a tap of a bell, a flick of the lights, or a clapping of hands.

Group responsibilities:

- Person "A" will get the tray of supplies for the group.
- Person "B" will begin the lab work by performing one test. The rest of the group will observe that person and record the results on individual tables of data. Then each person in the group will take a turn performing a test for the group. (This is particularly important with young and socially immature students who always want to be first or who want to monopolize the supplies.)
- Person "C" will return the tray of supplies at cleanup time.
- Person "D" will collect the four lab papers in his or her group and will put them in the container labeled "Lab papers."

6. *Manage the Science Lab*

- Give step-by-step directions.
- Give all directions and lab assignments before distributing trays of supplies.
- Assign numbers to tables or each group of students.
- Direct the supply people from tables 1, 3, 5, and 7 to get the supplies for their tables. Then direct the supply people from tables 2, 4, 6, and 8 to pick up the supplies for their tables. (This keeps movement of supplies and students to a minimum.)
- After all supplies are distributed, sit in on each group; observe, encourage, be as quiet as possible, but ask helpful questions and monitor student behavior.
- Give a five-minute warning before cleanup time.
- Collect all materials and clean up before any discussion.
- Collect supplies and papers: "Those responsible for returning supplies and turning in papers at tables 1, 3, 5, and 7 do so now." When those students are seated, call for the rest of the supplies and papers.

7. *Develop Your Own Hierarchy of Discipline*

Thus far your system of management teaches students what is expected of them. But what do you do when someone doesn't meet expectations? Follow your own hierarchy of discipline; have a logical, prioritized plan to help all students stay on the assigned task and not interfere with their own learning or that of others.

It is important to realize that some students and maybe even some parents will challenge your authority at the beginning of every school year. You need to be prepared with your own plan of action based on your school's policy so that you can act and not react to resolve problems.

A hierarchy of discipline needs to:

• Be consistent with the basic human needs for belonging and self-esteem. When basic needs are threatened, the individual fights back and the problem escalates and becomes more difficult to resolve.

• Distinguish between minor and major disruptions. Minor disruptions are ones which can be dealt with in the classroom without disrupting the entire science class, such as a student not attending to the assigned task or refusing to share science equipment within the group. Major disruptions are ones that can bring a class to a halt, and usually require assistance from the principal or parents. These include fights between students and consistent refusal to complete assignments.

• Provide the teacher with a prioritized list of responses that will resolve problems with the least classroom disruption: the teacher tries step one. If that works, no more steps are taken. If it doesn't work, then the teacher proceeds to step two, and so on.

Dealing with Minor Disruptions

When a minor disruption occurs, you have to decide if it is worth stopping what you are doing to deal with it. Minor disruptions frequently build into more troublesome ones. Thus, they need to be given attention but with the least amount of disruption to the class. Think of what you could do to stop the unwanted behavior without granting the offender the "power" of bringing the class to a halt. If that solution works, great. Stop there in your hierarchy. If it doesn't, move to step two, and so on. Try nonverbal communication first. Consider the following hierarchy of discipline which has worked for me:

Begin with nonverbal communication

(1) Observe the behavior. Make a mental note to observe that student.

(2) Look at the offender. Establish eye contact.

(3) Walk toward the offender. Stop walking when the student stops the unwanted behavior.

(4) Stand next to or behind the offender.

(5) Put your hand on the offender's chair.

Proceed to verbal communication

(6) Ask, "Do you need help?" Quietly resolve the problem.

(7) Direct the student to take work to the quiet table. If the student is removed from a lab experience, give him or her a choice after five minutes: return to the group or remain alone.

(8) Direct the student to write an explanation of the disruption and what should be done.

Make a file on each disruptive student

(9) Write a short description of the problem. File it along with the student's paper.

(10) Give the student a realistic choice, "Will you get to work or do I have to phone your parents to tell them that you are not completing your assignments?" Be prepared for the typical response, "I don't remember the phone number."

"I have home, work, and emergency phone numbers. Do you choose to get back to work? Or do I have to call your parents to discuss this situation?"

(11) If the child still doesn't cooperate, call and discuss the problem behavior with his or her parents. Document the conversation, and put it in the student's file. If you agreed to contact the parents within a certain period of time, make sure that you do so.

Dealing with Major Disruptions

(12) Send the file to the office, with either the problem student, or another child.

(13) Set up a conference with a parent, the principal, and the student to resolve the conflict.

(14) Continue to document and follow school policy for disruptive students.

A Personal Perspective on Discipline

It has been my experience that there is at least one student out of ten who does not respond to the nonverbal part of the hierarchy of discipline during the excitement of the hands-on activity. This child usually responds to:

• Being removed from the group and directed to a quiet or isolated desk within the class. (It is important for the student to see what he or she is missing. Most children want to be with their friends and enjoy the hands-on activities.)

• Being given a choice, after five minutes, of returning to the group and following all of the safety rules, or staying in the quiet area for the remainder of the class.

It has also been my experience that one out of ten students will need help learning how to work in a group. That one student will work his or her way down the entire hierarchy of discipline by the second or third week of school. But with emphasis on interesting scientific investigations, all students participate by the end of September or beginning of October. After that, the class appears to run itself and it is a joy to teach. All goes well until thoughts of Christmas vacation "dance in their heads." Then, as before each vacation, it is essential to provide lessons that are highly structured and highly motivating in order to compete for student interest.

8. Cancel Hands-on Activities, Science Labs, When You Are Absent

Your management system now consists of helping your students understand expectations and following a consistent hierarchy of discipline. But, what happens when you are not there?

For safety reasons, you must make provisions to stop all labs when you are absent. To do so, set up a substitute teacher file and place it where the substitute will be sure to see it. The substitute file needs:

• A note of explanation stating that no labs or hands-on activities are to take place in your absence. In place of any labs listed in your plan book, the substitute is to . . . You need to be specific.

a. Use the science text. State the pages.

b. Show a filmstrip. Give title.

c. Use a specific worksheet.

d. Turn the lab into a teacher demonstration.

• Directions for setting up audio-visual equipment and which students to put in charge.

• A list of reliable students' names.

• A list of names of students who usually need close supervision.

• Any information that will assist the substitute teacher, such as a class schedule, list of your duties, and an explanation of expectations and consequences.

How Do You Get Started?

Begin with planning. Not just a lesson plan, but a step-by-step plan for implementing the discovery approach in your classroom. Even if you have had experiences with discovery teaching, your students probably have had little or no experience with discovery learning. Your first lessons need to be teacher-managed to enable you to:

- teach safety rules and procedures
- motivate and stimulate curiosity
- identify students who know something about science investigations and those who do not
- identify students who are cooperative and those who are not
- identify students who have special learning needs
- plant the seed: "This class could be interesting."

The Latin Square Explained

The answer to the Latin Square is: A starfish.
Here are two ways of viewing the pattern in this Latin Square.

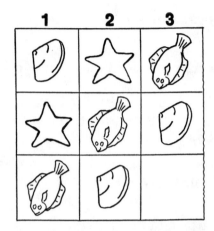

Figure 1-3. Answers to the Latin square

Columns 1 and 2 each contain a clam, starfish, and flounder; column 3 contains a flounder and a clam. It is missing a starfish.

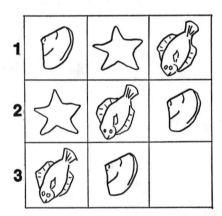

Figure 1-4. Answers to the Latin square

Rows 1 and 2 each have a clam, starfish, and a flounder; row 3 has a flounder and a clam. It is missing a starfish.

For more on Latin Squares and interesting problems from IQ tests see:

Lewis, David and Greene, James. *Thinking Better.* New York: Rawson, Wade Publishers, Inc. 1982.

For more on discovery learning:

Bruner, Jerome S. *The Process of Education.* Cambridge, Massachusetts: Harvard University Press, 1960.

Gagné, Robert M. *Conditions of Learning,* Second Edition. New York: Holt, Rinehart, and Winston, 1977.

Initiating Guided Discovery Science

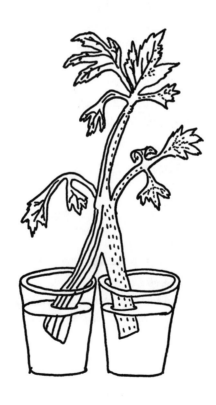

PART A
Get Ready

One way to initiate guided discovery science with hands-on experiences is to give students a problem to solve and provide them with a series of teacher-guided, exploratory experiences that will help them solve it.

An excellent problem to start with is to investigate the contents of a puzzle package. This investigation:

- does not require expensive scientific supplies
- is relatively easy to manage
- provides the teacher with a means of assessing student knowledge, logic, and behavior
- provides students with introductory experiences using the basic skills of science: observing, inferring, measuring, classifying, predicting, comparing, recording data, basing inferences and conclusions on data
- affords the teacher opportunities to teach the "whip," a multi-student response, and a "pass," a respectful way for a student to decline a teacher request to participate in a demonstration
- allows all students to participate
- can be adapted to a variety of grade levels
- can be adapted to a variety of science subjects by changing the contents of the packages

Lesson 1. A Demonstration: Investigating a Puzzle Package

OBJECTIVES

Students will:

1. Use three of the five senses to make observations.
2. Make an observation based on a change.
3. Make at least one measurement in both English and metric units.
4. Base inferences on observations.

MATERIALS

A ruler
Puzzle package: an opaque plastic bag containing a metal spoon
An identical, opaque plastic bag, empty (optional)
Scale (optional)

INSTRUCTIONAL PROCEDURES

Arrange the classroom so that everyone will have a view of the demonstration, and so that you will be able to carry the package to every student. Hold the bag for all to see and say, "We are going to investigate a puzzle package. Please do not call out, because it interferes with thinking. Think about what we could do that would help us learn about the contents of this package without opening it. Raise your hand when you have an idea that you would like to try." Quietly wait for three to five seconds before recognizing a student.

The most common responses from my students are, "Feel it" or "Hold it up to the light."

Follow up on a student response. Such as, "Would you like to feel the package without opening it?" If the student agrees, take the package to him. If not, ask for a volunteer.

Say, "Don't tell us what you think is in the package. Describe what it feels like."

Even though these directions are specific, some students jump to conclusions. Two frequent responses are:

"I don't know what's in the package."

"It's a . . ." The student names a specific object.

If you receive one of these answers, do not make a judgmental statement. Instead, rephrase the directions. Say, "We don't need to know what you think is in the bag, yet. We'll get to that later when we have more evidence. Tell us how it feels. Is it heavy or light? Soft or hard? Or it feels like a . . . ? **Note:** "It is a spoon," is an *inference*. "It feels like a spoon" is an *observation*.

After one person makes an observation, tell the class that you are going to bring the package to a few people so that they can also touch it. Say, "If you want to feel it, you may do so; but do not tell your observation until I point to you. If you do not want to touch it, just say, "Pass," and I'll move on to someone else."

A **pass** needs no teacher comment. Students are free to pass. For many, their curiosity is too great to pass. However, some may need to test you to see if you really mean what you say.

After five or six students feel the package, explain that you are going to do a **whip.** Say, "I am going to whip my arm around the room; when I point to you, tell us your observation or pass." As with anything new and different, a "whip" can make some students feel awkward at first. After a few trials, students realize that they really can pass (without a penalty) and do not have to give perfect responses every time to participate and eventually learn from the experience. It is an excellent way to involve a large number of students in a short amount of time and in a nonthreatening manner.

Continue to guide students to gather more information about the contents of the bag. Do not make positive or negative comments about the student responses because students respond to these sanctions either by trying to be right or by trying to get attention rather than thinking. Ask students to rephrase responses in order to turn statements of inferences into statements of observations.

Continue to gather information about the contents. Some examples:

- Shake it and listen.
- Tap it on a desk and listen.
- Approximate the measurement of its length.
- Weigh the puzzle package. Weigh an identical empty bag. Subtract the weight of the empty bag from the weight of the puzzle package to calculate the weight of the object.
- Feel the outline of the object.
- Press the plastic bag so as to reveal the outline of the object.

After all students have participated and enough evidence has been gathered to make logical conclusions, have students make inferences and think about how confident they are with their conclusions. To do so:

1. Review all of the observations. They might be:

- It feels hard.
- It weighs 1 ounce.
- It feels like a spoon.
- It is about six and a half inches long.
- It sounds like metal when tapped on the desk.
- It makes the outline of a spoon.

2. Then say, "Raise your hand if you think you know what is in the package." Wait for student response.

3. Say, "Raise your hand if you have no idea what is in the bag." Wait for student response.

4. Explain that an acceptable inference is one that makes sense according to the observations and data gathered. Thus, it is possible to have more than one acceptable inference.

5. Do a whip to elicit an inference, or pass, from each student.

Most students infer that it is a spoon. But if someone makes another acceptable inference be sure to explain why it is acceptable; for example, the inference "a bent piece of metal 6½ inches long" is supported by the data, the information gathered.

6. Next, ask, "Do you think scientists ever investigate something that they cannot see?" Listen to responses. Explain that scientists cannot see many things that they investigate such as the wind, electricity, and the inside of an atom. Consequently they must rely on the best possible information that they can gather. The

more reliable the information is, the more confidence they have in being able to reach a valid conclusion.

7. Say, "Raise your hand if you feel confident about your inference regarding the contents of this package?" Wait. Make no response.

8. Say, "Raise your hand if you are so confident that you do not need to look inside this bag."

Since this is the first lesson of this type for your class, it is recommended that you open the bag and show the spoon. Many students need to see what is in the bag to build confidence in their own ability to reach a valid conclusion.

If you have students who need to be told that they are right, you must wean them away from that need. To begin the weaning process, give an option.

9. Say, "I am going to open the package and show the object. If you want to look to check your inference, do so. If you do not want to look, don't." Open the bag and show the spoon.

Follow-up discussion questions:

- How do doctors, dentists, and veterinarians make observations?
- Do you want them to make observations without "opening up" the patient?

PART B
Get Set

Lesson 2. Investigating a Puzzle Package and Taking Notes Like a Scientist

OBJECTIVES

Students will:

1. Use at least three senses to make observations.
2. Make at least one measurement in both English and metric units. (No conversion.)
3. Record data on a table.
4. Make an inference by drawing a model that is supported by the data.

MATERIALS

Ruler

Puzzle Package A: A paper lunch bag containing a plastic prescription bottle big enough so that a small jingle bell placed inside it can roll and jingle.

Worksheet 2-1, "Puzzling Packages" (one for each pair of students)

ON THE BOARD

The information contained on Worksheet 2-1, which is a table of data.

INSTRUCTIONAL PROCEDURES

Demonstration

1. Explain that the purposes of this lesson are to investigate and take notes in a scientific manner by:

- investigating another puzzle package
- recording observations and measurements on a table of organized data
- recording at least one inference by sketching a model of what is believed to be the contents of the package, based on the observations

2. Point to the table of data on the board. Explain that you will record observations on the board while students record observations on their individual tables of data.

Worksheet 2-1

Name _____ Date _____

Puzzling Packages

Package	Observations	Measurements	Inferences	Sketch
A	1 2 3 4 5 6	1 2	1 2	
B	1 2 3 4 5 6	1 2	1 2	
C				
D				

3. Distribute Worksheet 2-1. Direct students to, "Write your name on the worksheet and then put the pencil down to signal that you are ready to begin the investigation."

4. Hold the package up. Say, "Raise your hand when you are ready to make an observation."

Possible response, "Hold it up to the light."

5. Explain, "I'm going to bring the package to a few people. You may pass or hold it up to a light. Do not say anything until I whip my arm around and point to you."

6. Take the package to five students. Then use the whip to elicit their observations.

7. Record the observations on the board. Direct students to record the observations on their individual tables of data and to check what they write, with what you write on the board. Explain that they do not have to use the same words but should have the same ideas.

8. Continue to:

- have students make observations
- revise statements of inference to statements of observation
- have students record information on their individual tables of data and check their information with that on the board.

9. Review all observations when you have sufficient data for students to make an inference about the contents of the package. These might be:

a. It's light.
b. It's round.
c. It's shaped like a can or a log.
d. It measures approximately 1 inch by 3 inches.
e. It makes a sound when shaken.
f. When the container is tilted, something inside it rolls and jingles.
g. Whatever is inside the can-like container, doesn't fill it.

10. Direct students to use this information to draw a model of what they infer or think is inside the package. You may hear some moans or comments, such as "I can't draw."

11. Reassure students that you are not interested in art work and they do not have to make a drawing of the exact contents.

12. Direct students to make a sketch of a model that makes sense based on the information gathered. Explain that the drawing will be acceptable if it is supported by the data, or not acceptable if it is not supported by the data. Remind them that more than one inference can be acceptable.

13. After most students have completed their drawings, access their confidence in their drawings. Say, "I'm going to do a whip to have you think about how confident you are in your drawing. Use a scale from zero to 10. Ten represents the most confidence; zero represents no confidence. When I point to you tell us the value you place on your confidence, or pass." Whip your arm around the class, include every student.

14. Identify a few students who are confident in their drawing. Ask a few volunteers to draw their sketches on the board and explain how they arrived at their ideas.

15. Discuss:

- observations that support the drawing
- the idea that scientists learn to work with the best possible evidence to develop inferences, hypotheses, and theories; and keep in mind that these carefully thought-out theories will need to be altered as the evidence dictates.

16. Introduce group responsibilities for hands-on experiences; say, "Notice that the worksheet has a space for Puzzle Packages "A" through "D." You are going to use this worksheet in the next science class in which you are going to work in groups of four. Each group will receive two puzzle packages to investigate. You will work with the people in your group to gather information. Each person will record the information on his or her table of data and will make an inference by describing and sketching a model of the contents."

17. Teach how to collect lab papers. Direct students to:

- Count off by fours.
- Those who are "fours" collect the papers from the people at their group and hold them.

18. Assign each table a number or letter. Direct those who are responsible for turning in papers at tables 1, 3, 5, and 7 to put the papers in the lab paper basket.

19. When those students are seated, direct those responsible for papers at tables 2, 4, 6, and 8 to put the papers in the basket.

20. Explain that you expect everyone to do well during the first science lab, a science class in which they use the science equipment to carry out an investigation, because they are well prepared. They now know how to:

- make observations, measurements, and inferences
- record information on a table of organized data
- make a sketch of a system based on observations.

PART C
Go!

Lesson 3. Initiating a Hands-on Experience with Puzzle Packages

OBJECTIVES

Students will:

1. Work cooperatively in a group.
2. Record observations and measurements on a table of data.
3. Make inferences based on the data.
4. Follow teacher directions for acquiring and returning science equipment.
5. Follow all safety rules and procedures or leave the group to write an explanation of the behavior problem and possible solutions.

MATERIALS

Prepare a tray of supplies for each group of students, containing:

> Ruler
>
> Magnet
>
> Package B: a bag containing a milk carton that contains one marble
>
> Package C: a box containing a magnet

Also prepare two packages for any groups that finish early:

> Package D: two paper clips in a box
>
> Package E: a nonbreakable salt shaker in a bag (Morton's® individual salt shaker)

Return Worksheets 2-1 to students.

ON THE BOARD

The four individual assignments as listed in the teacher's directions, or a modified version that meets the needs of your class.

INSTRUCTIONAL PROCEDURES

Directions

1. Give all directions and individual assignments before giving out papers and supplies. Have a student explain the directions in his or her own words.

2. Give individual assignments to students. Have them count off by fours.

- "Ones" will pick up the tray of supplies for the group.
- "Twos" will start the investigation for the group by selecting a package, performing one test while the group members watch. Everyone will write an observation. Then another person will perform a test on the package and everyone will observe and record those observations.
- "Threes" will return the tray at cleanup time.
- "Fours" will collect lab papers for the group and put them in the basket for lab papers.

3. Explain that they may discuss ideas quietly and may help each other with spelling, but that each person is responsible for his or her own paper. Papers will be graded individually on how well directions are followed and if the sketch of the contents of the package is supported by the data.

4. Say, "Your job in this lab is to work with the people in your group to make observations, record information on the table of data, and make at least one sketch showing what you infer is in the package. If you finish early, you may make more than one drawing for each package and/or you may raise your hand and ask for another puzzle package.

5. Say, "For safety reasons it is important that I can stop the lab at any time. If I blink the lights, stop what you are doing, look at me, and listen to the directions."

6. Select two students to return worksheets while you distribute supplies.

7. Distribute the trays of supplies. Say, "Supply people at tables 1, 3, 5, and 7 pick up your supplies. Each group may begin when it has a tray." Wait until those students are seated.

Say, "Supply people at tables 2, 4, 6, and 8 pick up your supplies."

8. When they are seated:

- walk to each group
- monitor their work and behavior
- use the nonverbal part of your hierarchy of discipline
- talk to any student who is disrupting a group or not following safety guidelines
- remove any student from the group who will not cooperate after you have spoken to him or her; follow your hierarchy of discipline
- be as quiet but as helpful as possible
- five minutes before cleanup time, give a warning that time is almost up.

9. Blink the lights to get students' attention. Direct cleanup. Say, "It is now cleanup time. People responsible for papers and trays at tables 1, 3, 5, and 7 turn them in now." When those students are seated, say, "Tables 2, 4, 6, and 8 turn in your papers and trays."

10. The follow-up discussion:

- Ask, "What do you infer was in Puzzle Package B?" Do a whip. Listen to their inferences.
- Ask for a volunteer to sketch an inference on the board and explain how the data supports the sketch.
- Ask, "What do you infer is in Puzzle Package C?" Do a whip.
- Ask for volunteers to put their sketches on the board and discuss how the data supports the sketch.

TEACHING TIME

The initial hands-on experiences usually require 45 minutes. The discussion can be conducted during the next science class. It can easily take 10 minutes to go over directions and distribute supplies and another 10 minutes to clean up. It is important to provide students with time to learn the lab procedures and science skills that will be used throughout the school year. They will improve with practice. **Note:** A lesson, including the hands-on experience, does not need to be completed in one class period.

Consider the Adaptability of Puzzle Packages

Puzzle packages are used by high school chemistry teachers who have their students investigate and draw a model of the unseen contents of a package. They then relate that skill to the investigations and thinking that scientists use to draw models of an atom. Elementary teachers use puzzle packages to teach the skills of observation and inference, and information about the senses. You can adapt the contents to meet the needs of your grade level and required content objectives.

PART D
Continue Discovery Activities

Lessons 1 through 3 are designed to initiate hands-on activities. They can also serve as lead-up lessons for a unit of study. In Lesson 3, Package B is a lunch bag containing a milk carton which has a marble in it. Package C is a box containing a bar magnet. This can lead to classifying objects that are and are not magnetic.

A Lead-up Lesson for a Hands-on Experience or Lab

Lesson 4. Identifying and Classifying Objects That Are and Are Not Magnetic

OBJECTIVES

Students will:

1. Use a ruler to make a copy of the table of data that is on the board.
2. Use a magnet to identify objects that are magnetic and not magnetic.
3. Classify objects as magnetic or not magnetic and record the information on the table of data.

MATERIALS

Rulers

A magnet

Recommended items to test:

Magnetic	Not Magnetic
Steel paper clips	Crayons
Iron nails	Wooden popsicle stick
	Copper wire

ON THE BOARD

Classifying Objects That Are Magnetic and Not Magnetic

Magnetic	Not Magnetic

INSTRUCTIONAL PROCEDURE

Demonstration

Say, "Look at the equipment. Raise your hand when you can tell us how to determine which items are magnetic and which are not." (Touch each item with a magnet.)

Direct students to copy the classification chart on the board and to signal that they are ready to begin the investigation by putting pencils and rulers down.

Have students make predictions.

Then invite a student to test an item and record the result on the classification chart on the board.

Direct all students to observe and record the result on individual worksheets and check their table with the board. Continue to test all items and record results.

At the end of the class, tell the students that in the next science class, they will work in lab groups to test and classify other items that are magnetic and not magnetic.

A Science Lab: A Hands-on Investigation

Lesson 5. Science Lab: Classifying Metal Objects as Magnetic and Not Magnetic

OBJECTIVES

Students will:

1. Construct and use a table of data.
2. Work cooperatively in lab groups.
3. Classify objects as magnetic or not magnetic based on test results.

MATERIALS

Rulers

On each tray of supplies:

Recommended objects to test. Identify the makeup of each item:

Magnetic	Not Magnetic
Iron hardware	Aluminum hardware
Steel paper clips	Aluminum paper clips
	Aluminum keys
	Copper hardware

ON THE BOARD

Classifying Metal Objects

Magnetic	Not Magnetic

INSTRUCTIONAL PROCEDURE

Direct students to make a copy of the classification chart on the board.

Because of Lesson 4, most students know what to do. Ask for a volunteer to explain the directions for the lab. After the directions are reviewed:

- Distribute the trays of supplies.
- Monitor each group, each student.
- Follow your hierarchy of discipline.
- Be a quiet observer in each group.
- Give a five-minute warning before cleanup time.
- Blink the lights to get students' attention.
- Give directions for cleanup.
- Identify objects that were magnetic and those that were not magnetic.
- Discuss what was learned from the lab.

For follow-up lessons on magnetism see Section 7. A study of magnetism is an excellent topic to investigate when first initiating hands-on experiences because:

- there are few safety concerns
- you don't have to be concerned about liquid spills
- it can lead to other investigations and other content areas such as electricity
- students can discover the law of magnetism and then experience the thrill of reading about one of their discoveries being a scientific law

PART E
Consider Different Contents
for the Puzzle Packages

Do you have a favorite unit of study; an area of scientific expertise? Think about how you could develop it into your own series of teacher-guided lessons to initiate guided discovery experiences. Devise your own:

- puzzle packages
- demonstration
- hands-on exploratory activities

PART F
Similar Approach,
Same Skills, Different Content

To help you think about developing your own set of teacher-guided lessons, here is another set of lessons that could be used to implement the guided discovery approach and to introduce an investigation of green plants. Use the teaching techniques previously described in Lessons 1-3, but use parts of plants in puzzle packages. Plastic bags are recommended because of the possible moisture in the plant parts.

MATERIALS

Package A: a white potato as an example of a root.

Package B: a leaf from any plant.

Package C: a stalk of celery as an example of a stem.

Package D: a large pod (lima or pea) as an example of a seed pod.

Package E: corn or sunflower seeds.

Possible Follow-up, Hands-on Investigations

Lesson 6. Investigating Leaves

OBJECTIVES

Students will:

1. Distribute trays of supplies in the usual manner, but will work individually.
2. Record observations and measurements on a table of data.
3. Make at least three leaf rubbings, each showing the veins and the edges of the leaves.
4. Follow teacher directions for distributing and collecting science supplies.

MATERIALS

Paper and pencil for notes and leaf rubbings

Paper and ruler to make a table of data

On each tray of supplies:

      Two to five leaves per student

      Magnifying lenses (optional)

ON THE BOARD

Investigating Leaves

Leaf	Observations	Measurements	Rubbings
A	Color Feel Smell Vein pattern Edge pattern Other	Width Length	
B			
C			

INSTRUCTIONAL PROCEDURES

Direct students to:

1. Use a ruler to copy the table of data from the board.
2. Observe a leaf carefully and record all of the information listed on the table of data.
3. Make a leaf rubbing by placing scrap paper under the leaf, new paper on top, and gently rub the side of pencil point over the leaf to reveal all of the veins

and the edges of the leaf. Label the rubbing with the letter that matches the description on the table of data.

After giving all of the directions:

- Distribute all supplies.
- Monitor the class.
- Give a five-minute warning before cleanup.
- Blink the lights to get students' attention.
- Direct cleanup.
- Discuss observations and measurements.
- Discuss types of vein patterns.
- Say, "Do you think that water gets into all of the leaves on a plant? How about in a giant tree? In our next classes, we will investigate the movement of water in plants that have leaves."

Lesson 7. Investigating How Water Moves Through Plants with Leaves and Stems

OBJECTIVES:

Students will:

1. Set up an investigation by following directions.
2. Record notes including observations, predictions, inferences, and diagrams.
3. Clean up spills as they occur, then return to the investigation.
4. Distribute trays of supplies in the usual manner, but work in teams of two within the group.
5. Write at least two other ways to investigate the movement of water in plants.
6. (For older students) Follow safety rules for using and passing a scalpel. Place the scalpel in front of the person who is ready to receive it. Scalpels, knives, and scissors are not passed hand to hand.

MATERIALS

One stalk of celery, with leaves, for every pair of students

For teachers with younger students: a sharp knife, kept in a protective covering

On each tray of supplies:

   Four clear plastic cups, each with a piece of masking tape on its side for the child's name

A large plastic beaker or jar of water

A container of red and blue food coloring

For older students: two dissecting trays and two scalpels

Teaching Tip

Purchase celery that is in bunches, has leaves, and has some roots at the base of the stalk. Keep it in water or in a refrigerator until ready to use.

ON THE BOARD

List Student Directions:

Investigation: How Does Water Move Through a Plant with Leaves?

1. Sketch the Celery Investigation. Label everything.
2. List what you predict will happen in this investigation.
3. List some other ways that we could investigate the movement of water in plants.

INSTRUCTIONAL PROCEDURES

Say, "Look at the supplies. What can we do with them to observe the movement of water in a plant with leaves?" If no one suggests placing a plant in water with food coloring, suggest it.

Demonstrate how to set up Investigation 1:

1. Write your name on two jars.
2. Pour the same level of water in each jar.
3. Add five to ten drops of red food coloring in one jar and the same amount of blue food coloring in the other jar.
4. Cut the celery to remove the base containing the roots and the lower part of the stems. Set aside for Investigation 2.
5. Then cut the stalk part of the way up the middle from base to leaves.
6. Place one part in the red water, the other in the blue water. See Figure 2-1.

Continue with the rest of the directions:

7. Refer to the notes on the board. Say, "After you have set up the celery leaf investigation, follow the directions on the board. Sketch a diagram to show what you have done. Label everything in the diagram. Under your sketch, list your predictions. That is, write what you think will happen to the celery."

After going over all of the directions:

• Distribute the supplies.
• Monitor the class.

red   **blue**

Figure 2-1. Celery in water with red and blue food coloring. Red–solid lines. Blue–dotted lines

- Give a five-minute warning before cleanup time.
- Blink lights to get students' attention.
- Give cleanup directions.
- Direct students to place plants in sunlight or under artificial light.

Set up Investigation 2:

- Place less than half an inch of water in a jar.
- Add food coloring.
- Place the discarded base of the celery, roots down into the water. (See Figure 2-2.)
- Place where students can observe.

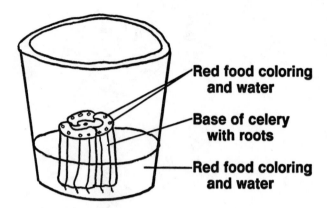

Red food coloring
and water

Base of celery
with roots

Red food coloring
and water

Figure 2-2. The base of a celery plant in water with red food coloring

Note: The height, condition of the celery, and light determine the rate at which the water rises in the celery. It takes about 12 hours for the dyed water to reach the leaves.

ALTERNATIVES OR ADDITIONS

1. Set up one container with red water and one with clear water. Identify the celery in the clear water as one of the controls in the investigation, along with temperature and light conditions. Identify the celery in the dyed water as the experimental part of the investigation.

2. Direct the groups to conduct different investigations with the celery.

Lesson 8. What Happened to the Celery? A Close Look at the Vascular System of a Plant with Leaves and Stems

OBJECTIVES

Students will:

1. Observe and compare the parts of the celery that were in the red and blue colored water.
2. Observe and sketch a fresh cut in the stem.
3. Break the stem (by aiming it down into a paper towel in order to keep water and dye from splashing), and pull the pieces apart to examine the tubes.
4. Observe the base of the celery in dyed water. See Figure 2-2.
5. Write an inference to explain how water gets to the leaves in celery plants.
6. Write an inference to explain how water enters and moves in plants with stems and leaves, including giant trees.

MATERIALS

For older students: scalpels and dissecting trays

For teachers with younger students: a sharp knife

Students' notes from last lesson

On each tray of supplies:

Two paper towels

Students' celery investigation

Red and blue pencils or crayons

Magnifying glass (optional)

ON THE BOARD

List Student Directions:

1. Sketch a diagram showing the results.
2. Diagram a cross-section of the celery stem. See Figure 2-3.
3. Observe the roots and stems in Investigation 2.
4. Write an explanation of how you think water moves in the celery.
5. Write an explanation of how you think water travels through other plants with leaves, including tall trees.

INSTRUCTIONAL PROCEDURES

- Review the directions on the board.
- Distribute the celery investigations.
- Monitor each group.
- Take Investigation 2 to each group for observations.
- Give a five-minute warning before cleanup.
- Blink the lights to get students' attention.
- Give directions for cleanup.
- Ask for observations and inferences.

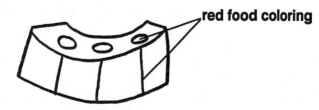

Figure 2-3. A cross-section of celery that has been placed in water with red food coloring

Follow-up Discussion

    1. Discuss the vascular system of the plants that have leaves. Water enters the plant through its many small roots. The water travels up the tubes to the veins in the leaves.

    2. Explain that plant scientists classify plants as vascular and nonvascular. Celery is an example of a vascular plant. It has tubes which transport or carry water to the leaves. Some other examples of vascular plants are trees and flowering plants, including those that produce fruits and vegetables. Nonvascular plants do not have tubes to transport water through the plants. Examples are mushrooms, molds, algae, and lichen.

Teaching Tips

- A sharp knife helps lessen the possibility of damaging and clogging the tubes in the celery.
- Make all cuts for young or immature students.
- Ask a florist to donate some white flowers for investigations of the vascular system of flowering plants.
- Set up an investigation with a nonvascular plant such as store-bought mushrooms.
- Invite students to set up a similar investigation at home, with parents' consent.

Oh No, Not with This Class!

    If you are currently teaching a science class, you may be thinking that water, dye, scalpels, and snapping wet plants are just not the kinds of "ingredients" that your students can handle properly. If you are right, then you should not teach Lessons 7 and 8 exactly as written.

    If only a few students cause behavior and safety problems, remove them from the group and give them a highly structured assignment. Continue to invite them to participate if they feel they are ready to follow directions and safety rules. If most of your students are not ready to work safely in a lab type, hands-on activity, modify the lessons to meet the needs of your class. Consider:

    1. Turning the lessons into demonstrations.

    2. Modifying the lessons. Give the directions for Lesson 6, Investigating Leaves. While most students are observing leaves and making leaf rubbings, supervise one group of students as they set up Celery Investigation 1. Place yourself so that you can see the "celery" group and the rest of the class at the same time. It will still require two or three lessons to complete all of these investigations.

Teacher Guidance and Student Practice

Students improve with guidance and practice. You improve their chances for success when you:

- Stimulate their interest.
- Refrain from making judgments about their thinking.
- Make your expectations clear.
- Demonstrate that you expect quality work from everyone.
- Treat everyone with respect and in a consistent manner.

For those who have learning problems, especially with reading and traditional academics, you can provide opportunities for success while participating in quality experiences. For some it may be their first success with academics.

Personal Perspective on Guided Discovery and Discipline

Everyone appreciates the feelings and respect that come with success, achievement, and quality. Most students would rather do something at which they can succeed. Even those who are conditioned to disrupting a class, usually prefer to be busily engaged in an interesting and challenging task. Teacher concern about potential discipline problems is a major deterrent to guided discovery science. However, efficiently managed, guided discovery lessons have solved discipline problems for me.

After the first three weeks of school, my in-class discipline problems decrease to almost none. Unfortunately, my severe discipline problem students perform elsewhere; in the cafeteria, on the playground, anywhere there is less structure and guidance. But they don't like it if I am informed of their poor behavior. They don't want me to lose my "high regard" for them which has developed as a result of their work in science class. I cannot solve all of my students' problems but by using the interest and excitement of science activities presented in an environment which is carefully managed, I can teach them in my own classroom and all of my students can learn.

Through effectively managed, interesting, guided discovery science lessons, you can create a stimulating and humane environment in which you can teach and your students can learn and sometimes soar.

Working and Thinking as Scientists

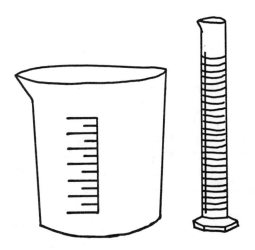

Personal Perspective

Few of your students will become scientists. Some will not attend college and some will not even graduate from high school, but all are and will be continually confronted with problems and the need to make decisions. You can introduce them all to lifetime skills and positive attitudes that lead to sound decision making and problem solving by teaching them to think and work as scientists.

To begin, provide your students with a problem and help them develop a safe approach to gathering and analyzing data before making a decision based on that data. The conclusions can range from an easy to understand answer to the concept that there is insufficient data to solve a problem. Students need to learn that all problems do not have easy answers and that even scientists do not have enough data to answer all questions.

In the discussion part of your lessons, ask your students to identify several ways in which a problem can be investigated. This will help them to develop an understanding that there is more than one way to investigate a problem and to help them avoid the misconception that there is only *one* scientific method.

Stress applying the new information and skills to decision making, both in and out of the classroom. Discuss the importance of thinking and working as a scientist in making personal decisions and solving problems. For example, after teaching the effects pollution and smoking have on the lungs, you can point out that while *some* air pollution problems are beyond the control of students, others are not. For instance, only they can decide if they are going to smoke or not.

Students Love to Work Like Scientists

Learning to calculate an average doesn't sound exciting to most adults. However, when children are permitted to do an activity that will help them solve this problem, they love it, and are not shy about saying so.

One visitor who came to our fourth grade science class was surprised to see elementary students using graduated cylinders and eyedroppers to gather data and then calculate the mode (value or number that occurs most frequently) and median (the middle answer when all answers are arranged in order) to determine the average number of drops in a milliliter. He was invited to participate. He asked some students, "Do you like science?"

"Yes."

"Why?"

"We get to do neat things like finding modes and medians."

"Do you really like that stuff?" He couldn't believe his ears.

"Yea!" was the response from the four boys in his group.

They really do "like that stuff," because they get to use the eyedroppers and graduated cylinders, and they can make the calculations.

However, they would not love finding the mode and median if it were taught strictly as a paper and pencil calculation. Ask any adult who was taught averages or statistics in the abstract. He or she usually thinks that the mean (the sum of the answers divided by the sum of the observations) is the only average and rarely loves the calculation. The visitor who came to class had met many people who had been taught calculations without the hands-on activities. No wonder he was surprised when he met children who enjoyed their school work.

Important Steps to Teaching Problem Solving and Statistics

To teach students to work and think in a scientific manner, the following steps are helpful:

1. Provide all of the background information that is necessary to solve the problem.
2. Begin with a simple problem that you think every student in your class can understand and solve.
3. During initial lessons, provide specific directions and procedures. As students demonstrate that they can gather and interpret data, encourage them to develop their own methods of procedures and tables of data.
4. Proceed to more complex problems by building on the concepts and skills that you are teaching.
5. Teach for student success by providing class time for introduction, practice, built-in reviews, and mastery.

Teaching Tip

Substitute teachers should not conduct science labs. The substitute may not be a science teacher and some students do not behave as well for substitutes as they do for their own teachers. The worksheets, that do *not* require hands-on activities, make excellent lessons to leave for substitutes.

Using the Tools of a Scientist

Lesson 1. Introducing the Graduated Cylinder

OBJECTIVES

The students will:

1. Use a graduated cylinder to measure the amount of liquid in a container in milliliters.
2. Record the measurement as a number and unit of measurement, for example, 2 ml.
3. Read the bottom of the meniscus, the dip in a liquid, if there is one.

MATERIALS

A variety of different size graduated cylinders
A 1000 ml beaker or jar of water with food coloring

INSTRUCTIONAL PROCEDURES

Introduction

Hold up a graduated cylinder that is graduated by ones. Ask if anyone knows what it is. Most elementary and junior high school students do not know because they have not had any experience with graduated cylinders. However, you may find a few students who have chemistry sets at home and can explain their use. If you do not, explain that it is used to measure the volume in milliliters of substances that pour.

Demonstration

Pour some colored water into the cylinder up to the one milliliter mark to demonstrate the small size of a milliliter and the use of the markings on the cylinder. Tell the students that there is one milliliter of water in the cylinder and that each mark on this cylinder represents one milliliter. Add one more milliliter to the cylinder and ask, "How many milliliters are now in the cylinder?" (Two.)

Ask, "If I were to add three more milliliters of water, at which mark do you predict the water level would reach?" (Five.)

Add three more milliliters and observe that the water mark rises to the 5 ml mark.

Actually Reading a Graduated Cylinder

Explain that it is easy to be off by a milliliter or two by looking up or down on a cylinder or by holding it on a tilt. To avoid this:

1. Place the cylinder on the desk or flat surface.
2. Put your eye level to the water level.
3. If there is a dip in the water level, read the bottom of the dip. The scientific term for this dip is the meniscus.

Next, add some more water to the cylinder. Have another student read the number of milliliters of water that are in the cylinder using the new guidelines, and record the amount and units on the board.

Note: Water in glass graduated cylinders is more likely to form a convex meniscus, or dip, than water in plastic cylinders.

Introduce an "If You Like" Assignment

Consider giving an "If You Like" assignment which is an invitation, not a requirement. It is evaluated and treated as extra credit. "If You Like," find examples of measurements in milliliters in everyday life by reading food labels, and present them in the form of a montage, poster, or mobile.

Problem Solving with the Graduated Cylinder

Lesson 2. How Many Milliliters of Water Are in Each Jar?

OBJECTIVES

Students will:

1. Use graduated cylinders to measure the number of milliliters of water in each of four jars.
2. Record the measurement as a number and a unit of measurement.

MATERIALS

For the demonstration:
    Jar A containing 20 ml of colored water
    A 25 ml graduated cylinder
For the lab, place the following items on each tray of supplies:
    Graduated cylinders
    Four jars with different amounts of colored water:

Jar B - 5ml
Jar C - 10ml
Jar D - 15ml
Jar E - 25ml
Paper towels
Worksheet 3-1, "How Many Milliliters of Water Are in Each Jar?"

ON THE BOARD

Copy the worksheet.

INSTRUCTIONAL PROCEDURE

Hold up Jar A. Ask, "Using the supplies on the tray, how can we find out how many milliliters of colored water are in this jar?" (Pour the water into a graduated cylinder.)

Demonstration

Pour the contents of Jar A into a graduated cylinder, put your eye level to the water level, read the number of milliliters, and record the measurement on the table of data on the board.

Jar A contains 20 ml of colored water. Record that on the board to demonstrate how students should record information on the worksheet.

Specific Lab Directions

Direct students to:

1. Measure the contents of each jar.
2. Record the measurements on the worksheet.

For those who finish early:

3. Practice measuring the contents of each jar again and recording the measurements next to B^2, C^2, and so on.
4. Compare the first and second measurements of each jar to determine if they are the same. Explain any discrepancies. (Spills, lack of skill in measuring.)

Before beginning the lab, review the general lab procedures by asking for a volunteer to name the four lab jobs. The four jobs in each group are:

1. Pick up the tray of supplies.
2. Get the lab work started.
3. Return the tray at cleanup time.
4. Collect lab papers and put them in the science basket at cleanup time.

STUDENT WORKSHEET 3-1

Name _____   Date _____

How Many Milliliters of Water Are in Each Jar?

Jar	Number of Milliliters/Unit of Measure
A	_____ ml
B	_____ _____
C	_____ _____
D	_____ _____
E	_____ _____
B^2	_____ _____
C^2	_____ _____
D^2	_____ _____
E^2	_____ _____

After an orderly cleanup, discuss acceptable measurements. Make allowances for spills. If time permits, ask students to identify some of the measurements in milliliters that they have noticed on containers.

Lesson 3. What Is the Capacity of Each Container in Milliliters?

OBJECTIVES

Students will:

1. Record the measurements from the labels of at least three containers.
2. Measure the number of milliliters of water that each container holds when filled to capacity.
3. Write at least one inference about the relationship between the reported measurements listed on the label and the actual measurement when the container is filled to capacity.

MATERIALS

On each supply tray:

    A graduated cylinder
    An empty jar
    A pie tin
    A large beaker of water with food coloring
    Paper towels

Suggested containers:

    Juice container
    Milk carton
    Soda can
    Vinegar bottle

Suggested bonus containers:

    Oil bottle
    Shampoo bottle
    Moisturizing lotion bottle
    Eyedrop container

Student Worksheet 3-2, "What Is the Capacity of Each Container in Milliliters?"

Name _____ Date _____

What Is the Capacity of Each Container in Milliliters?

Container	No. of ml recorded on the container	No. of ml of water when filled	Inference
Juice container			
Milk carton			
Soda can			
Vinegar bottle			
Bonus containers			

INSTRUCTIONAL PROCEDURES

Hold up a container. Ask, "How can we find out how many milliliters this container will hold? (Read the label. Measure it.)

Ask, "How can we determine how many milliliters of water this container will hold when filled to capacity?" (Fill it with water and use a graduated cylinder to measure the number of milliliters of water.)

Lab Directions

Direct students to:

1. Read and record the number of milliliters on a label.
2. Determine the amount of milliliters of water the container will hold when filled to capacity. Record that number in the proper column.
3. Compare the measurement recorded on the label with the actual measurement. Are they the same? Or are they different?
4. Write at least one inference that would account for the comparisons.
5. If a group finishes early, the supply person may pick up a couple of bonus containers and the group can determine their capacities.

DISCUSSION

After cleanup, discuss the inferences that would account for any discrepancies in the amount recorded on the container and the amount the container holds when filled to capacity. Possible inferences and points for discussion:

1. The container company doesn't know how to measure.
    a. A company whose personnel cannot measure will not be in business long.
2. Companies do not fill all containers to capacity because:
    a. Some products need room for expansion and contraction.
    b. Some products handle better when not filled to capacity.
    c. Many containers have a mark or a ridge that indicates the level to which the container is filled.
    d. Some are filled to capacity.
3. The measurement on the container indicates the amount of the product in the container.

Lessons That Lead to an Understanding of Comparisons and Averages

Lesson 4. How Many Drops of Water Are in a Milliliter?[1]

OBJECTIVES

Students will:

1. Complete at least four trials to determine the number of drops in a milliliter.
2. Record the number of drops in each milliliter on a table of organized data.
3. Write a conclusion about the number of drops in one milliliter.

MATERIALS

On each tray place:
   One eyedropper
   A large jar of water with food coloring
   A graduated cylinder graduated by ones
   Paper towels
Student Worksheet 3-3, "How Many Drops of Water Are in One Milliliter?"

INSTRUCTIONAL PROCEDURES

Say, "How could we find the number of drops of water that are in one milliliter? Look at the 'clues' on this tray of supplies."

If someone has an acceptable answer, have that person demonstrate it. If not, demonstrate one way to determine the number of drops of water in one milliliter by:

1. Filling the eyedropper with colored water.
2. Holding the dropper perpendicular to the desk surface.
3. Counting the number of drops that it takes to fill the container to the one milliliter mark.

Specific Lab Directions: Explain Multiple Trials

After distributing the worksheets, direct your students to each take a turn squeezing drops of water into a graduated cylinder in order to determine the

---

[1] *Science—A Process Approach/Part D.* "Measuring 14, Measuring Drop by Drop. American Association for the Advancement of Science/Xerox Corporation, 1968.

Name _____ **Date** _____

How Many Drops of Water Are in One Milliliter?

Trial	Number of Drops	Conclusion
1		
2		
3		
4		
5		
6		
7		

number of drops in one milliliter. The first time that is accomplished is called trial one, and the number of drops is recorded on the table of data next to trial one. Everyone in the group records the number of drops for each trial on his or her own table of data. After four trials are recorded, everyone looks at the data and writes a conclusion about the number of drops in a milliliter. Any group that finishes early should complete more trials and determine whether or not the additional data supports the conclusion.

Teaching Tip

Since many elementary students do not know how to use a dropper, you will need to demonstrate how to squeeze the air out of the dropper before any water can go in it, as well as how to hold it perpendicular to the desk surface.

DISCUSSION

1. Did every trial result in the same number of drops? (No)
2. Why do you think this happened?
3. What conclusion did you make about the number of drops in one milliliter?

Some acceptable responses:

1. Every trial did not result in the same amount of drops.
2. Different eyedroppers make different size drops.
3. Different people make different size drops.
4. It is difficult to exert the same pressure on the eyedropper in making each drop.

Explain that in situations such as this, when every trial does not necessarily result in the same number, scientists and mathematicians look for averages and other numbers that provide useful information. The next few labs will explore identifying useful numbers.

Lesson 5. Introducing the Mode

OBJECTIVES

Students will:

1. Define the mode as the value or number that occurs most frequently in a set of numbers.
2. Identify the mode in a set of numbers.

MATERIALS

Worksheet 3-4, "Calculating the Mode"

INSTRUCTIONAL PROCEDURES

To introduce the mode, discuss the difficulty encountered during the previous lesson when trying to answer the question, "How many drops are in a milliliter?" Explain that there are ways of organizing and tabulating a series of numbers so that one or two significant numbers can be identified that are useful in answering the question.

Explain that one way to analyze the data is to identify the number that occurs most frequently. This observation is called the mode.

Demonstrate How to Identify the Mode

List the following numbers on the board: 30, 29, 29, 31. Ask, "If the number of drops for each trial is: 30, 29, 29, and 31 respectively, what is the mode? That is, what is the number that occurs most frequently?" (The mode is 29 because it is the number that occurs most frequently.)

Ask, "Would the mode be any different if I rearrange the order of the observations or measurements?" (No)

Since this is a new concept, give an other example: 5, 5, 5, 3, 2, 5, 3. (The mode is 5 because it is the number which occurs most frequently.)

Distribute Worksheet 3-4, "Calculating the Mode." Direct the students to look at the table of data on the worksheet. Tell them to look at the number of drops, write the mode in the proper column, and put their pencils down to signal when finished.

After they are finished, discuss their answers and permit students to change their answers if necessary. Stress that this is part of learning a new skill, not cheating. (The mode for the table of data on the worksheet is 29 because it is the number that occurs most frequently.)

Next, direct your students to work independently and write the mode for each set of numbers on the worksheet.

After five or ten minutes, students should be ready to correct their work. Have a volunteer go to the board and explain a problem. Review each problem with the class.

Answers to the Problems on the Worksheet: "Calculating the Mode."

The mode on the table of data is 29.

Name _____ **Date** _____

Calculating the Mode

The **mode** is the value or number that occurs most frequently in a given series of numbers.
Directions: Write the mode for each series of numbers given.

Trial	Number of Drops in a Milliliter	Mode
1	29	
2	30	
3	29	
4	28	

Problems for further practice.

1. Numbers	Mode
28	
30	
30	
30	

2. Numbers	Mode
20	
15	
19	
22	

3. Numbers	Mode
300	
290	
300	
299	
300	

4. Numbers	Mode
5000	
3000	
3000	
5300	
5200	

Answers to Problems for Further Practice:

1. The mode is 30.
2. There is no mode because no number occurs more frequently than the others. This happens sometimes. The answer is none, *not* zero.
3. The mode is 300.
4. The mode is 3000.

Lesson 6. Identifying the Mode to Determine the Number of Drops in One Milliliter of Water.

OBJECTIVES

Students will:

1. Identify the mode as one type of average.
2. Determine the number of drops in one milliliter of water based on their lab data.
3. Calculate the mode from a list of numbers.

MATERIALS

On each tray:
>   One eyedropper
>   Jar of water with food coloring
>   Graduated cylinders
>   Paper towels

Worksheet 3-5, "How Many Drops Are in One Milliliter of Water?"

INSTRUCTIONAL PROCEDURES

Before the lab, review:

1. The meaning of mode; the number that appears the most frequently in a series.
2. The concept that each group will not get the same number for the mode because of all the variables that cannot be controlled.

Lab Directions:

Complete as many trials as possible in the class time and identify the mode for those trials.

STUDENT WORKSHEET 3-5

Name _____ Date _____

How Many Drops Are in One Milliliter of Water?

Trial	Number of Drops	Mode	Conclusion
1			
2			
3			
4			
5			
6			
7			
8			

© 1989 by The Center for Applied Research in Education

Lesson 7. Introducing the Median

OBJECTIVES

Students will:

1. Identify the median as one type of average.
2. Define the median as the middle value in a set of numbers that have been ranked according to value.
3. Calculate the median from a list of numbers.

MATERIALS

Worksheet 3-6, "Calculating the Mode and the Median"

ON THE BOARD

How Much Does a Pack of Gum Cost?

Brand	Cost in Cents	Mode Most Frequent	Median Midpoint	Conclusion
1	$.29			
2	$.28			
3	$.30			
4	$.25			
5	$.28			

INSTRUCTIONAL PROCEDURES

Point out that different stores frequently charge different prices for the same item. Explain that one way to compare values is to survey the price of an item at several stores and determine the cost most frequently charged and the median price.

Direct the students' attention to the table of data written on the board. Have a volunteer identify, write, and explain the mode for this series of numbers. (The mode or the most frequently charged price for the gum surveyed is 28 cents.)

Introduce the Median

Explain that another way to identify a useful number from a series of numbers is to identify the number that falls in the middle. To do this, the values need to be

STUDENT WORKSHEET 3-6

Name _____   Date _____

Calculating the Mode and the Median

The **mode** is the number that occurs most frequently in a series of numbers. The median is the middle number in a series of numbers arranged in order of value. There are three steps to identifying the median in a series of numbers:

1. Write the numbers in order of value.
2. Continue to cross out the top and the bottom numbers until the middle number is identified.
3. Circle the middle number.

What is the Cost of a Fully Equipped, Five-Speed Sports Car?

Brand	Cost in Dollars	Mode Most Frequent	Median Midpoint	Conclusion
A	30,000			
B	40,000			
C	18,000			
D	30,000			
E	29,000			

Practice finding the mode and median.

1. Numbers	Mode	Median
20		
20		
20		
19		
21		

2. Numbers	Mode	Median
500		
600		
400		
700		
300		

placed in order of size, so that the middle value can be determined. The term for this middle value, or midpoint, is the median. There are three steps in identifying the median:

1. Arrange the numbers in order of value.
2. Cross out the top and the bottom numbers. Repeat this step until the midpoint is identified.
3. Circle the midpoint.

Demonstrate How to Find the Median

"How Much Does a Pack of Gum Cost?"

Step 1. Rewrite the numbers in order of value.

25
28
28
29
30

Step 2. Cross out the top and the bottom numbers to find the mid-point.

~~25~~
~~28~~
28
~~29~~
~~30~~

Step 3. Circle the middle number.

~~25~~
~~28~~
(28)
~~29~~
~~30~~

Independent Practice Finding the Mode and Median

Distribute Student Worksheet 3-6, "Calculating the Mode and Median." Read the directions to the class. Direct students to work independently. They are to write the answers for the mode and median, to show all of the steps in identifying the median for each problem, and to be prepared to discuss each problem.

When students are finished, work the problems out on the board and tell students to correct their papers and make any changes that are necessary so that the paper will be a helpful reference during the next lab.

Answers to the Worksheet: "Calculating the Mode and the Median"

What Is the Cost of a Fully Equipped, Five-Speed Sports Car?

Brand	Cost in Dollars	Mode Most Frequent	Median Midpoint	Conclusion
A	30,000	30,000	~~18,000~~	
B	40,000		~~20,000~~	The most frequent cost for the cars surveyed is $30,000.
C	18,000		(30,000)	
D	30,000		~~30,000~~	
E	29,000		~~40,000~~	The median cost is $30,000.

Answers to "Practice Finding the Mode and Median"

1. Numbers	Mode	Median
20	20	19
20		20
20		20
19		20
21		21

2. Numbers	Mode	Median
500	None	300
600		400
400		500
700		600
300		700

Lesson 8. Finding the Mode and Median

OBJECTIVES

Students will find the "average" cost of a bicycle by calculating the mode and median costs.

MATERIALS

As many catalogs as possible, at least two per group.

INSTRUCTIONAL PROCEDURES

Hold up a catalog and ask "How can we find the price most frequently charged for a bicycle and the median cost of bicycles?"

Your students should be able to identify a method of gathering and analyzing data. Their method should include the idea of looking up the cost of bicycles in the catalog, recording those costs on a table of data, and then identifying the mode and median. They may need some help in organization, such as:

1. Making the table of data.
2. Identifying one person in each group who will trade catalogs with other groups.
3. Deciding on a reasonable number of prices to list based on class time.

Once you and your students have determined an acceptable method of investigation, direct them to make their own tables of data and distribute the catalogs.

DISCUSSION

After all of the data are gathered and analyzed, discuss the findings. Identify:

1. The most frequently charged price for a bicycle.
2. The median cost of a bicycle.
3. The types of bicycles surveyed.

Teaching Time: One to two days; more if you survey other items.

Lesson 9. Introducing the Mean

OBJECTIVES

Students will define the mean as the "average" which is obtained by dividing the sum of two or more quantities by the number of those quantities.

MATERIALS

Worksheet 3-7, "Calculating the Mode, the Median, and the Mean"
Note: Since the mean requires division as well as addition, it can only be understood by those students who have a working knowledge of simple division. If your students have not mastered simple division, skip Lessons 9 and 10, and teach Lesson 11 or delete calculating the mean.

Name _____   Date _____

Calculating the Mode, the Median, and the Mean

To calculate the mean:

1. Add all of the numbers together.
2. Count the quantity of numbers in the series.
3. Divide the sum of the numbers by the quantity of numbers.

What Is This Student's Average Based on These Grades?

Grades	Mode	Median	Mean	Conclusion
90				
92				
89				
95				
89				

Additional Practice

1. Numbers	Mode	Median	Mean
20			
20			
22			
25			
20			
22			

2. Numbers	Mode	Median	Mean
100			
101			
100			
101			
100			
100			

Answers to the Worksheet: "Calculating the Mode, the Median, and the Mean"

What Is This Student's Average Based on These Grades?

Grades	Mode	Median	Mean	Conclusion
90	89	~~89~~	$\dfrac{455}{5}=$	This student has a median of 90 and a mean of 91.
92		~~89~~		
89		(90)	91	What grade would you assign this student? Why?
95		~~92~~		
89		~~95~~		

Additional Practice

1. Numbers	Mode	Median	Mean
20	20	~~20~~	$\dfrac{129}{6}=$
20		~~20~~	
22		(20	21.5
25		22)	or
20		~~22~~	22
22		~~25~~	

Additional Discussion for This Problem

When there is an even number of values and you want to find the median, what do you do? Find the mean for the two middle numbers. That is, add the two middle numbers and divide their sum by 2.

In this example, twenty and twenty-two are circled in the middle. Find their mean:

$$\begin{array}{r} 20 \\ +\ 22 \\ \hline 42 \end{array} \qquad 42 \div 2 = 21$$

The median is 21.

2. Numbers	Mode	Median	Mean
100	100	~~100~~	$\dfrac{602}{6}=$
101		~~100~~	
100		(100	100.3
101		100)	or
100		~~101~~	100
100		~~101~~	

INSTRUCTIONAL PROCEDURES

Introduce the Mean

Explain that the median and the mean are arithmetic averages but when most people speak about an average, they usually refer to the mean.

The most common way that teachers compute student grades is to compute the mean.

Distribute Worksheet 3-7, "Calculating the Mode, the Median, and the Mean." Read the directions to the class. Demonstrate how to calculate the mean of the grades listed on the worksheet:

- The sum of the grades is 455.
- There are 5 grades.
- The mean is arrived at by dividing 455 by 5.
- The mean is 91.

Direct students to complete the worksheet and be prepared to discuss their answers.

When students are finished, correct the answers by working the problems out on the board and having students correct their own papers.

Teaching Time: One to Two Days.

Lesson 10. How Many Peas in a Pod?

OBJECTIVES

Students will:

1. State the average number of peas in a pod based on lab work.
2. Identify the mode, median, and mean.

MATERIALS

On each tray:
    At least two pods per student
    Paper towels
    Student Worksheet 3-8, "How Many Peas in a Pod?"

Lab Directions

Direct students to:

Name _____   Date _____

How Many Peas in a Pod?

Pod	Number of Peas	Mode	Median	Mean	Conclusion
1					
2					
3					
4					
5					
6					
7					
8					

1. Count the number of peas in each pod.
2. Calculate the mode, median, and mean.
3. Write a conclusion identifying the average number of peas in a pod.
4. Show all calculations on the worksheet.

Teaching Tip

Since your students have previous lab experience, permit different groups to approach the problem differently. One group might prefer to have one person open all of the pods, especially in older groups of students. Younger students usually demand a turn, or think it unfair if they don't get a turn.

Lesson 11. A Different View of Your Findings: Compute the Range

OBJECTIVES

Students will:

1. Define the range as the difference between the highest and lowest values.
2. Identify the range as an evaluation tool.

MATERIALS

Worksheet 3-9, "Analyzing and Deciding How to Report the Data"

INSTRUCTIONAL PROCEDURES

Explain that another way to view the data is to look at the two extreme observations rather than at a typical observation or average. To do this, identify the highest and the lowest values. Then, subtract the lowest from the highest value. The difference is known as the range. It can be written as:

Range = Highest − Lowest or
$$R = H - L$$

Using the Range

In investigations where you can expect the observations to be similar, such as the number of drops in a milliliter, the lower the difference, meaning the lower the range, the more confidence you can place in the results. The higher the range, the less confidence you can place in the data. You must then consider:

Name _____   Date _____

Analyzing and Deciding How to Report the Data

How Many Hours Will a 100-Watt Light Bulb Work?

Bulb	Working Hours	Mode	Median	Mean	Range	Report
1	200					
2	199					
3	200					
4	202					
5	201					

What Was the Temperature Today?

Readings	Temp in °F	Mode	Median	Mean	Range	Report
5:00 AM	60					
7:00 AM	65					
9:00 AM	70					
11:00 AM	80					
1:00 PM	89					
3:00 PM	90					
5:00 PM	92					
7:00 PM	85					
9:00 PM	75					

1. Carefully review the procedures used.
2. Examine the values to determine if one of the values is "out of line" with the rest of the data. If so, it may be discounted. Example: one low grade from an excellent student.
3. Consider redoing the lab work.

In some investigations we cannot expect the observations to be close in value. This latter type of reporting data is most commonly used by the United States Weather Bureau. Daily temperatures are reported as the high, low, and median temperatures. These three values give a better understanding of the daily temperatures than any one value.

Directions

Direct students to work out the problems on the worksheet and to use the range to help write a brief report in which they:

1. State if they have a high or low degree of confidence in reporting the results as an average. Or . . .
2. Report the results as the high, low, and the median or mean.
3. Compare answers with someone who is finished.

After fifteen to twenty minutes, work the problems out on the board, and have students correct their own papers.

Answers to the Worksheet: "Analyzing and Deciding How to Report the Data"

How Many Hours Will a 100-Watt Light Bulb Work?

Bulb	Working Hours	Mode	Median	Mean	Range	Report
1	200	200	~~199~~	$\dfrac{1002}{5} =$	$\dfrac{\begin{matrix}202\\-199\end{matrix}}{3}$	A high degree of confidence can be placed in the average of 200 working hours.
2	199		~~200~~			
3	200		(200)	200.4		
4	202		~~201~~	or		
5	201		~~202~~	200		

What Was the Temperature Today?

Readings	Temp in °F	Mode	Median	Mean	Range	Report
5:00 AM	60	None	~~60~~	$\frac{706}{9} =$	92 − 60	The high was 92°F.
7:00 AM	65		~~65~~		32	
9:00 AM	70		~~70~~	78.4		The low was 60°F.
11:00 AM	80		~~75~~	or		
1:00 PM	89		(80)	78		The median was 80°F.
3:00 PM	90		~~85~~			
5:00 PM	92		~~89~~			The "average" temp was 78°F.
7:00 PM	85		~~90~~			
9:00 PM	75		~~92~~			

If there is time, discuss other examples of averages and ranges that students have noticed since they started this study. Discuss the possible uses for businesses, such as:

- A shoe store has to know what size shoes to stock.
- An ice cream store must identify the most frequently ordered flavors of ice cream, and so forth.

Lesson 12. Putting It All Together

OBJECTIVES

Students will design a procedure to answer the question, "How long does it take a LIFE SAVER® to melt?"

MATERIALS

Classroom clock (optional with second hand)

On each tray:

A piece of waxed paper containing one LIFE SAVER® per student

Student Worksheet 3-10, "How Long Does It Take a LIFE SAVER® to Melt?"

Note: Students who are not to have sugar may pass eating the LIFE SAVER®, but will complete the worksheet.

Name _____ **Date** _____

How Long Does It Take A LIFE SAVER® to Melt?

LIFE SAVER®	Dissolving Time in Minutes	Mode	Median	Mean	Range
1					
2					
3					
4					
5					
6					
7					
8					

INSTRUCTIONAL PROCEDURES

Ask your students how they can use all of their new skills and information to determine the number of minutes that it takes LIFE SAVERS® to melt? Permit different groups to develop different procedures. Distribute the supplies and work-sheets. When everyone is finished, discuss the results.

Additional Questions for Investigation

For further labs gathering and analyzing data to identify the mode, median, mean, and range, consider:

1. How many petals are on a buttercup?
2. How many seeds in _____? You decide on the fruit or vegetable. Maybe make a salad and enjoy it.
3. How many teeth on a comb?
4. How many "eyes" in a pair of sneakers?
5. What is the average cost of a pair of nondesigner jeans?
6. What is the average cost of a pair of designer jeans?
7. How much television does a person watch in one day?
8. How long does the flavor last in a stick of gum?

INDEPENDENT STUDY

For an independent study or science fair project, encourage students to identify their own interests, design a safe study, gather data, analyze that data to identify the mode, median, mean, and range. Then prepare a brief report.

SECTION **4**

Introduction to Chemical and Physical Changes

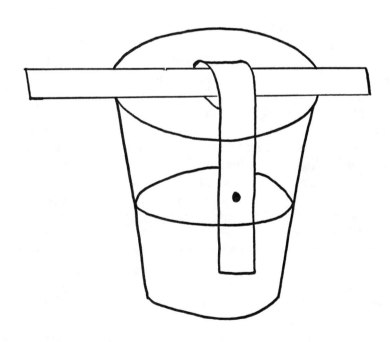

Personal Perspective

Children observe changes early in their lives. They ask many questions to get some information about the changes that they see but don't understand, such as, "Why are leaves green?" or "Why do the leaves change color?" The chemistry of plants holds the answers to these questions. Chemistry also holds the answer to many other questions, such as, "What happens to the water in a puddle when the puddle disappears?" It is a shame that so many nonchemists fear, rather than understand and enjoy, the everyday chemistry in our lives.

As an elementary teacher, you are not in the business of training chemists, but you should be providing opportunities for successful encounters with easy to understand physical and chemical changes. Early success with beginning chemistry, and an introduction to its importance in our daily lives, helps promote positive attitudes and can lessen the fear of chemistry later in the students' lives.

Introducing students to beginning physical and chemical changes actually solved a serious problem for me . . . "What to do with overenthusiastic children during the two weeks before the Christmas/Winter Break?" During this time, the students' ability to concentrate is "inversely proportional" to the nearness of the vacation. Some get so excited that they become physically ill. Some lose self-control and get into numerous arguments and fights.

I discovered, serendipitously, that elementary age students love chemistry with its interesting names, symbols, and surprises of new products produced by chemical change. I was actually looking for a fail-safe project that everyone could complete successfully during the last two days before our winter vacation—something that they could make and enjoy during their vacation.

For years, during my winter vacations, I had enjoyed making, eating, and giving gifts of stuffed dates. I decided to try it with my students. They loved the stuffing, but only a few liked the dates. Simple modifications and a little creativity took care of that. My students in grades three through seven, and those in my university science methods classes have enjoyed making these and other candies.

Because our study of chemistry comes just before Christmas, I am frequently asked by parents if I recommend chemistry sets for children. I tell them that I usually do *not* recommend chemistry sets for most elementary age children because some of the chemicals in the sets can be dangerous; and because the directions require careful reading. However, if they are willing to work with the student, then chemistry sets can provide hours of enrichment. But I follow that with a true story about a friend that took place during his formative years.

His mother was sitting in the living room when, for no apparent reason, she began to cry. At first she didn't know what was wrong. Then she yelled, "Bob, are you using your chemistry set?" Yes he was. He had just learned to make a tear gas. Bob is now a chemist.

To teach an introduction to physical and chemical changes, you need not have majored in chemistry. However, you must understand the differences between physical and chemical changes, and you must stress safety to your students.

Most of my students have the misconception that chemistry is "mixing stuff" together to see what happens. You must explain to your students that due to chemical changes, which result in a different and potentially hazardous substance, mixing can be dangerous and they are not to do it. When they get older and study chemistry in high school or college, they will have more opportunities to experiment. For now, they must *not* "mix stuff."

They should concentrate on understanding the difference between physical and chemical changes, and on identifying products and processes in our lives that exist because of chemistry. They also should think about how much they like chemistry. It might provide an interesting and rewarding career for them in the future.

Lesson 1. Introduction to Physical and Chemical Changes

OBJECTIVES

Students will:

1. Define a physical change as a change in the appearance of a substance with the actual substance remaining the same.
2. Define a chemical change as a change from one substance to a new substance or substances.
3. Distinguish between a physical and chemical change.

MATERIALS

Two sheets of paper
One jar of water
Metal tongs
Heat-resistant surface
Matches

INSTRUCTIONAL PROCEDURES

Demonstration One/Roll and Burn Paper

1. Hold up a piece of paper. Roll it up into a ball. Ask, "Is this still paper?" (Yes)

2. Explain that it *is* still paper, but it is changed in appearance. Chemists call this type of change a physical change.

3. Hold up another piece of paper. Tear it in half. Ask, "Is it still paper?" (Yes) "What kind of change?" (Physical)

4. Next, work over a nonflammable tray. Use standard safety procedures to light a match, close cover before striking and strike the match away from you. Hold one half of the paper with tongs. Light the paper. Place it on the tray. Use the tongs to pick up the charred remains. Ask, "Is it still paper?" (No)

Explanation

The paper is no longer paper. It has been chemically changed into carbon and some gases that rose in the air with the smoke. Chemists call this a chemical change.

Demonstration Two/Wet Paper

1. Take the other half of the paper and dip it in a jar of water. Hold up the wet paper, ask, "Is it still paper?" (Yes) "What kind of change?" (Physical)

2. Write the classification chart, as shown in Figure 4-1, on the board.

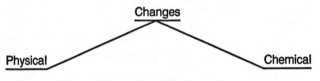

Figure 4-1. A classification scheme

MAKING A CLASSIFICATION CHART

Direct students to copy the chart and to list the four changes that you demonstrated under the correct heading.

Discuss the changes, list them on the board and direct students to make any changes in their notes that are necessary. See Figure 4-2. Classifying Physical and Chemical Changes.

Tell students that chemical changes can be dangerous because they use energy and produce new products which can be hazardous. For these reasons, they are not

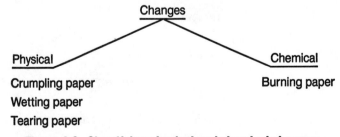

Figure 4-2. Classifying physical and chemical changes

to experiment or play with chemicals. However, if they want to try some chemistry at home, they can do some of the "If You Like" assignments that will be recommended during the study of chemistry.

Stress Safety

Stress that in chemistry safety is so important that the chemical industry spends millions of dollars a year on safety programs and procedures. If the best chemists and other scientists are concerned about safety, we must realize how important safety rules are, and follow them both in and out of class.

Lesson 2. The Chemistry of a Burning Candle

OBJECTIVES

Students will:

1. State that melting wax is a physical change.
2. State that burning wax and wick are chemical changes.
3. State that burning chemically turns substances into carbon, gases, and water vapor.

MATERIALS

Fire-resistant surface, such as a frozen dinner tray

Candle in a holder or melted to a jar lid

Matches

Jar which will fit over the candle

INSTRUCTIONAL PROCEDURES

Demonstration

1. Have a few students feel inside the jar and describe what they feel. (It is dry and empty.)

2. Explain the safety procedures for striking a match as you close the cover before striking the match away from you. Reemphasize the importance of safety in studying chemistry. Have the students observe the burning candle. (The flame is yellow. The wick is black. Smoke is rising. Wax is melting.)

3. Ask, "What type of a change takes place when wax melts?" (Physical) "Is it still wax?" (Yes)

Explanation

The wax melts, drips down the side of the candle, forms a puddle, and hardens as it cools; physical changes take place. However, all of the wax doesn't melt. Other changes take place when a candle burns.

4. Tell the class that you are going to put the jar over the candle. Ask for predictions.

5. Put the jar over the candle. The candle goes out, smoke rises, and water vapor forms on the inside of the jar. Explain that burning chemically changes things into: carbon, seen as the black on the top of the jar, and gases, which rise with the smoke, and something else.

6. Lightly touch the jar to make sure that it is cool. When it is safe to touch, have a few students feel inside the jar and make an observation. (The jar feels wet.)

Burning chemically produces water that is given off in small amounts as water vapor.

Explanation

When the wax melts, all of it does not drip down the side of the candle. Some of it is absorbed by the wick and travels to the top of the wick. The wax and the wick burn. That is, the heat causes them to chemically combine with the oxygen in the air and make new products of carbon, gases, and water vapor.

PROBLEM SOLVING

Have students apply what they have learned by asking the following questions.

1. What kind of change occurs when a candle wick burns? (Chemical)
2. When wax melts? (Physical)
3. When chocolate melts? (Physical)
4. When a log burns? (Chemical)
5. When anything burns? (Chemical)
6. Burning chemically changes a substance into three substances. What are they? (Carbon, gases, and water vapor.)

Lesson 3. Physical Changes with Water. Is it Still Water?

OBJECTIVES

Students will:

1. Identify melting, freezing, and evaporating as examples of physical changes.
2. Identify ice, steam, and water as different forms of water.

3. State that there are three states of matter: solid, liquid, and gas.

4. State that the chemical formula for water is H_2O.

MATERIALS

A tray of ice cubes

A saucer

One jar of water

One tea kettle

A hot plate

A large metal cooking spoon, chilled, with an insulated handle

One tin can or ice tray for freezing some water

A Pyrex cup

Potholder

INSTRUCTIONAL PROCEDURES

Demonstration One/Water to Ice

Pour some water in a can or ice cube tray. Ask, "What do you predict will happen if we put this in a freezer?" Put it in a freezer and observe it during the next science class.

Demonstration Two/Ice to Water

Place an ice cube on a saucer. Ask, "What do you predict will happen if we let this ice cube sit out for a few days?" Listen to predictions. Place the saucer containing the ice cube in a safe place when it can be observed for the next few days. Observe and discuss the changes. (It melts; evaporates. Physical changes.)

Demonstration Three/Ice to Water to Steam

Put some ice in the tea kettle. Place it on a hot plate. As the ice is heating up, ask the students to predict what will happen to the water. When it turns to water, pour some in the Pyrex cup for the class to see. As the steam comes out of the spout, ask if it is still water.

Place the chilled bowl of the spoon in the steam as it leaves the spout of the kettle. Water will condense on the spoon. Ask:

1. "What do you observe?" (Water on the spoon.)

2. "What is happening to the water in the kettle?" (It is turning to steam.)

3. "What happens when the steam hits the spoon?" (It turns to water.) Introduce the term, "condenses." Scientists say that the steam "condenses" to form water.

4. "What kind of change is it when water changes to steam?" (Physical. It is still water.)

5. When water evaporates? (Physical. It is still water.)

6. When water freezes? (Physical. It is still water.)

Explanation

Water can exist in the three states of matter: solid, liquid, or gas. When water is a solid, it is called ice, but it is still water . . . solid water. Chemists write the chemical formula for water as H_2O. When it is a gas, it is called steam or water vapor. A change in the state of matter is a physical change. We physically change water for washing, cooking, and keeping foods and drinks cool.

Safety Concern

Changing water to steam is dangerous because steam can cause severe burns. Consequently, this should be a teacher demonstration.

Lesson 4. Ice, Water, and Steam Are Chemically the Same but Physically Different (Grades 6–8)

OBJECTIVES

Students will:

1. State that ice, water, and steam are chemically the same.

2. State that temperature determines if water will exist as a solid, liquid, or gas.

3. Make a molecular model of water to demonstrate the physical differences in ice, water, and steam.

MATERIALS

On each supply tray:

Water molecule patterns. See Figure 4-3.

Scissors, closed and pointed down in cans

One sheet of paper

INSTRUCTIONAL PROCEDURES

Hands-on Activity

Place the students in groups of four.

Direct students to:

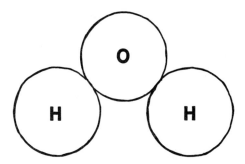

Figure 4-3. Water molecule pattern

- Take out a sheet of paper.
- Trace around the pattern.
- Cut it out.
- Label it H-O-H to match the letters on the pattern.

To signal that they are ready for the next set of directions, tell students to:

- Put the scissors, closed and pointed down, in the container.
- Put all of the scrap paper to one side of the table.
- Put the sheet of paper in the middle of the table.

When students are ready, ask them what they think their model represents. (Water)

Explanation

Explain that their model is one of a water molecule. A molecule is a group of atoms. Atoms are small, distinct, and separate units of matter. Their water molecule consists of two atoms of hydrogen and one atom of oxygen.

Directions for Hands-on Activity One/A Model of Ice

Tell students to:

- Place one sheet of paper in the middle of the table.
- Place one model of a water molecule in each corner of the paper.
- Make each water molecule vibrate by moving it in place.
- Leave the molecules in place and listen.

Explanation

They have made a model of ice. Molecules are always in motion. They are affected by temperature. When the temperature reaches 32°F, the freezing point for water, the water molecules cannot move about freely, but vibrate in place. Ice is solid water.

Directions for Hands-on Activity Two/A Model of Water

Tell students to:

- Make the molecules vibrate again.
- Move them about on the paper.
- Keep the molecules the same distance apart as before as they move them across the paper.

Explanation

When the temperature goes above freezing but stays under the boiling temperature, 212°F, the water molecules vibrate and move freely. Ice physically changes to water.

Directions for Hands-on Activity Three/A Model of Steam

Tell students to:

- Make the molecules vibrate.
- Move them freely about the paper.
- Move the molecules ten times farther apart than before by moving them upward off the paper.

Ask what they think this model represents. (Steam)

Explanation

When the temperature reaches the boiling point, the water molecules move rapidly throughout the container, and eventually leave the container, as the steam left the kettle in an earlier lesson.

Teaching Tip

Have the students repeat the molecular representations of these physical changes. This time have the steam condense into water, and then have the water freeze to ice. Have students explain each step.

Review

Water can exist as a solid, liquid, or a gas. Its chemical formula is H_2O. A change in the state of matter is a physical change.

Lesson 5. Indirectly Observing the Effects of Temperature on Molecular Movement

OBJECTIVES

Students will:

1. Demonstrate that the temperature of water affects the time required for food coloring to mix evenly with the water.
2. State that mixing water and food coloring is a physical change.

MATERIALS

A large, classroom clock with a minute and second hand

One copy per student of Student Worksheet 4-1, "Effects of Temperature on Mixing Water and Food Coloring"

On each supply tray:

One jar of hot, tap water

Same size jar, with the same amount of ice water

One container of food coloring

INSTRUCTIONAL PROCEDURES

Hands-on Experience, Lab

Distribute Student Worksheet 4-1, "Effects of Temperature on Mixing Water and Food Coloring." Direct each lab group to:

1. Gently place three to five drops of food coloring in the hot water.
2. Record observations.
3. Without stirring, measure and record the time, in minutes and seconds, that it takes for the two to mix evenly.
4. Repeat these procedures using a few drops of food coloring and ice water.
5. Write at least one inference about the effects of temperature on the movement of water molecules that would be explained by the observations.

Explanation

The higher the temperature, the faster the molecular movement. The lower the temperature, the slower the molecular movement. Food coloring takes longer to mix in the ice water than it does in the warm water. We cannot directly observe molecular movement but we can observe it indirectly by:

Name _____   Date _____

Effects of Temperature on Mixing Water and Food Coloring

Table 1: Observing Unassisted Mixing of Water and Food Coloring

	Food Coloring and Hot Water	
Observations	Mixing Time in Minutes & Seconds	Inferences

Table 2: Food Coloring and Ice Water

Observations	Mixing Time in Minutes & Seconds	Inferences

1. Watching the mixing of food coloring in different temperatures of water.
2. Pouring a strong-smelling liquid, such as vanilla extract, in a jar and measuring the time that it takes the odor to travel throughout the room.

Lesson 6. Indirectly Observing Spaces Between Molecules

OBJECTIVES

Students will:

1. State that 5 ml of water and 5 ml of sugar make less than 10 ml of sugar water.
2. Develop a hypothesis to account for 5 ml of water and 5 ml of sugar making less than 10 ml of sugar water.

MATERIALS

One copy per student of Student Worksheet 4-2, "Do 5 ml of Water and 5 ml of Sugar Make 10 ml of Sugar Water?"

On each supply tray:

A 10 ml graduated cylinder

A graduated medicine cup containing 5 ml of sugar

A jar of warm water

A stirrer

INSTRUCTIONAL PROCEDURES

Lab

Direct students to:

1. Predict the amount of sugar water that will result from mixing 5 ml of water with 5 ml of sugar.
2. Record the prediction on the worksheet.
3. Measure 5 ml of water in the graduated cylinder.
4. Add the 5 ml of sugar to the water, in the cylinder, and stir.
5. Measure and record the volume of sugar water.
6. Write a hypothesis that would explain the results.
7. Develop a test for the hypothesis.

Explanation

There is less than the expected amount of 10 ml of sugar water because there are spaces between the sugar crystals which are filled with air. There are also

Name _____ Date _____

Do 5 ml of Water and 5 ml of Sugar Make 10 ml of Sugar Water?

DIRECTIONS:

1. Think about mixing 5 ml of water with 5 ml of sugar. Would it equal 10 ml of sugar water? Record your prediction.

   Prediction _____

2. Place the 5 ml of water and 5 ml of sugar in the graduated cylinder. Stir until the sugar dissolves. Record the volume of sugar water in milliliters.

   Volume of sugar water _____

3. Write a hypothesis, an explanation that you can test, that would account for your measurement.

   Hypothesis _____

   _____

   _____

   _____

   _____

4. Explain how it would be possible to test your hypothesis.

   _____

   _____

   _____

   _____

   _____

   _____

   _____

   _____

   _____

   _____

spaces between the water molecules. When the sugar dissolves in the water, the sugar and water molecules get closer together. The water fills in the air spaces. Thus, mixing sugar and water in any amount results in a volume slightly less than the sum of the two volumes.

Teaching Tips

Match the unit of measurement to whatever units your class has been studying:

1. One cup of water plus ¼ cup of sugar equals less than 1¼ cups of sugar water.
2. Eight ounces of water plus 2 ounces of sugar equals less than 10 ounces of sugar water.

Consider having students test their hypothesis using different amounts of sugar and water.

Lesson 7. Elements, Mixtures, and Compounds

OBJECTIVES

Students will:

1. Identify iron and sulfur as elements.
2. Identify iron plus sulfur as a mixture.
3. Identify iron plus sulfur plus heat as a compound.
4. Write a definition of an element as a substance consisting of one kind of atom.
5. Write a definition of a mixture as a combination of substances in which the substances can be separated without a chemical change.
6. Write a definition of a compound as a substance composed of two or more elements, in fixed proportions, which can be separated only by a chemical change.

MATERIALS

Iron filings
Powdered sulfur
A paper plate
A plastic or wooden stirrer
Magnet in a plastic bag
Heat source: candle, alcohol lamp

Test tube

Test tube holder

Matches

A paper with a crease for pouring the mixture into the test tube

One copy per student of Student Worksheet 4-3, "Elements, Mixtures, and Compounds"

Heat-resistant plate

INSTRUCTIONAL PROCEDURE

Demonstration One/Physically Change Iron and Sulfur

1. Use a well ventilated room.

2. Hold up a container of sulfur. Pour some on a paper plate. Have students describe its physical properties. (Yellow. Powder. Smells like a match.)

3. Identify it as an element, sulfur. Invite a lab assistant to test for magnetism by using the magnet that is protected by the plastic bag. (It is not magnetic.)

4. Hold up a container of iron filings. Pour some on the paper plate. Do not let them mix with the sulfur. Have students describe its physical properties. (Gray-black. Looks like pepper. Makes a sound when poured.)

5. Identify it as an element, iron. Have another lab assistant test it for magnetism. (It is magnetic.)

6. Have someone use the stirrer to mix the iron and sulfur on the plate. Have the students describe its physical properties. (The colors of yellow and black blend to make a grayish yellow. Particles of sulfur and iron can be seen.)

7. Ask if the two substances can be separated. Have someone test the substances with the magnet. (It can be separated. The iron filings are attracted to the magnet.)

Ask:

a. Is the iron still iron? (Yes)

b. Is the sulfur still sulfur? (Yes)

c. What kind of change results when iron is mixed with sulfur? (Physical)

Explanation

Stirring the iron filings and the sulfur powder together changes them physically and makes a mixture. In a mixture the substances can be separated without a chemical change. In this mixture, each substance maintains its original properties. Thus, the iron remains magnetic and the sulfur remains nonmagnetic; and the mixture can be separated using a magnet.

Name _____   Date _____

Elements, Mixtures, and Compounds

DIRECTIONS: Circle the correct answer according to the demonstrations.

1. Iron is an example of (an element, a compound).

2. An element is a substance that is made of one kind of (atom, compound).

3. Iron filings are (magnetic, not magnetic).

4. Sulfur is another example of (a compound, an element).

5. Sulfur is (magnetic, not magnetic).

6. Sulfur mixed with iron makes (a mixture, an element).

7. A magnet (does or does not) attract iron filings in a mixture of iron and sulfur.

8. Making a mixture of iron and sulfur is an example of a (physical, chemical) change.

9. After heating the mixture of iron filings and sulfur, the magnet (does or does not) separate the iron filings from the mixture.

10. Heating the mixture of iron filings and sulfur makes a new substance named (iron sulfide, iron ore).

11. Chemists call this new substance a (mixture, compound).

12. Heating a mixture of iron filings and sulfur to make iron sulfide is an example of a (physical, chemical) change.

DIRECTIONS: Complete the definitions according to the information presented in the demonstrations.

Definitions:

A. An element is made of one kind of _____.

B. A mixture is a combination of substances which can be _____ without a chemical change.

C. A compound is a substance composed of two or more elements, in fixed amounts, which

   cannot be separated without a _____.

Demonstration Two/Chemically Change Iron and Sulfur

1. Make sure that the room is well ventilated.

2. By weight, mix the two parts iron with one part sulfur.

3. Use safety procedures to strike a match and light the heat source.

4. Place mixture on a piece of creased paper and pour the mixture into a test tube.

5. Heat the mixture in the test tube using standard safety practices and explain each step:

   - Use a test tube holder.

   - Slant the test tube in such a way that the opening is not pointed towards anyone.

   - Continuously move the test tube so that the tube and substances heat evenly.

6. When the sulfur and iron combine to make a silvery "molten" substance, pour it on the heat-resistant plate. Some of it will remain in the tube.

7. Ask if this substance can be separated by a magnet.

8. Invite a lab assistant to test it with a magnet. (The substance cannot be separated with a magnet.)

Explanation

Mixing iron and sulfur is an example of a physical change. The iron is still iron and the sulfur is still sulfur. This type of a blend is called a mixture. A mixture is a combination of substances which can be separated without a chemical change. This mixture can be separated by a magnet.

When a mixture of iron and sulfur is heated, a chemical change results. The new substance that results is named iron sulfide. Iron sulfide is a compound. A compound is made up of two or more substances which cannot be separated without a chemical change. Accordingly, the compound, iron sulfide, cannot be separated with a magnet.

Safety Concerns:

1. Due to the fumes given off when making iron sulfide, the room must be well ventilated.

2. Seat students with known respiratory and allergy problems away from the demonstration area and close to open windows or arrange for them to go to another teacher during the demonstration.

3. Use standard safety procedures when striking the match and lighting the heat source.

4. Use standard safety procedures for heating a substance in a test tube, as explained in the demonstration directions.

Teaching Tips

1. Use two or three class periods to accomplish the objectives of this lesson.
   - Day 1. Work through the demonstrations and the verbal explanations.
   - Day 2. Give each student a copy of the worksheet "Elements, Mixtures, and Compounds." Repeat the demonstrations and help the students fill in the information.
   - Day 3. If more time is needed, finish the worksheet. Correct the worksheets and return them to the students for use as notes.
2. Keep the magnet in a plastic bag. Iron filings make a mess of magnets.

Answers to Student Worksheet 4-3, "Elements, Mixtures, and Compounds"

1. element	5. not magnetic	9. does not
2. atom	6. mixture	10. iron sulfide
3. magnetic	7. does	11. compound
4. element	8. physical	12. chemical
A. atom	B. separated	C. chemical change

Lesson 8. Symbols, Formulas, and Equations

OBJECTIVES

Students will:

1. State the chemical symbols for sulfur (S), iron (Fe), hydrogen (H), and oxygen (O).
2. Write the chemical formula for iron sulfide (FeS).
3. Write the equation for making iron sulfide:

   $Fe + S \rightarrow FeS$

MATERIALS

A black gum drop
A yellow gum drop
One tooth pick

INSTRUCTIONAL PROCEDURES

Listen and Learn, Minilecture

Remind the students that they know some chemical symbols for elements. Ask for some examples. Tell them that chemists have a symbol for every element. There are currently 106 elements known. The symbols are usually the first or first and second letters of each element's name. Sometimes the English, Greek, or Latin names are used, so all symbols aren't what we expect. The chemical symbol for sulfur is S; and for iron is Fe. "Fe" comes from the Latin name for iron, *ferrum.* Print the chemical symbols on the board. Point out that the "F" is capitalized and the "e" is not.

Chemists combine symbols to write the chemical formulas of compounds. The formula for iron sulfide is FeS. Write it on the board.

Chemists write the symbols and formulas to explain chemical reactions in expressions called equations. Remind the students that they observed you chemically change iron and sulfur into iron sulfide by using heat. One way to help understand this is to make a model.

Demonstration

1. Hold up the yellow gum drop. Ask what it could represent. (Sulfur)
2. Hold up the black gum drop. Ask what it could represent. (Iron)
3. Join them together with a tooth pick. What does this represent? (Iron sulfide)

Writing Like a Chemist

Chemists use symbols and formulas to write an equation that represents what happens during a chemical change. Write on the board:

$Fe + S \rightarrow FeS$

Ask if someone would like to try to explain the equation or read it. (Iron and sulfur chemically make iron sulfide.)

"If You Like" Assignments

1. Write the names and symbols for as many elements as you can find.
2. Write the names and formulas for as many compounds as you can find.
3. Use gum drops, jelly beans, or clay, and tooth picks to make models of chemical formulas. Label each with a name and formula.

Lesson 9. Four Steps for Writing a Chemical Equation

OBJECTIVES

Students will write the steps for learning to write a chemical equation:

- Write the key words.
- Write the symbols under the words.
- Decide if the equation is balanced. (For grades 6 and up.)
- Write the balanced equation the way a chemist would.

MATERIALS

Teacher: chalkboard
Students: paper and pencil

INSTRUCTIONAL PROCEDURES

Listen, Learn, and Take Notes

Explain that they are going to use four steps to learn to write chemical equations. They will start by writing the equation for making iron sulfide. Direct them to copy whatever you write on the board.

Write the Four Steps to Writing an Equation

1. Write the key words:
    Iron   plus   sulfur   chemically   make   iron   sulfide
2. Write the symbols:
    Fe + S   →   FeS
3. Is it balanced?
    1 iron   1 iron
    1 sulfur   1 sulfur
    Yes, it is balanced.
4. Write the balanced equation.
    $Fe + S \rightarrow FeS$

Practice Using the Four Steps for Writing a Chemical Equation

Write the chemical equation for making water.

1. Hydrogen plus oxygen chemically make water
2. H_2 + O_2 $\rightarrow$ H_2O
3. Is it balanced?

   2 H   2 H
   2 O   1 O

   No, it is not balanced.
4. The balanced equation is:

   $2\,H_2 + O_2 \rightarrow 2\,H_2O$

Write the chemical equation for making hydrochloric acid.

1. Hydrogen + Chlorine $\rightarrow$ Hydrogen chloride
2. H_2 + Cl_2 $\rightarrow$ HCl
3. Is it balanced?

   2 H   1 H
   2 Cl   1 Cl

   No, it is not balanced.
4. The balanced equation is:

   $H_2 + Cl_2 \rightarrow 2\,HCl$

Teaching Tip

Explain how to balance an equation but do *not* require all students to learn how to balance equations. Rather, extend it as an invitation or a challenge. Those who are ready for it will take up the challenge and enjoy it. Those who are not will be spared the frustrations of being pushed beyond their mathematical and abstract learning limits.

Liken a balanced equation to a baking recipe which gives the amount of each ingredient to use and the number of servings it yields.

Lesson 10. Fun with Formulas (Grades 5–8)

OBJECTIVES

The students will:

1. Write chemical formulas when given the chemical makeup of a compound.

2. List the chemical makeup of a compound when given a chemical formula.

MATERIALS

A list of formulas:

Water H_2O

Iron Sulfide FeS

Glucose $C_{12}H_{22}O_{11}$—table sugar

Baking soda $NaHCO_3$

Sulfuric acid H_2SO_4

Iron Oxide or rust Fe_2O_3

Student Worksheet 4-4, "Fun with Formulas"

INSTRUCTIONAL PROCEDURES

Listen and Learn, Minilecture

Ask, "What is the formula for iron sulfide?" (FeS)

Write it on the board. Explain that the chemical formula identifies the elements and the number of atoms of each element. According to this formula, there is one atom of iron and one atom of sulfur in iron sulfide.

Explain: If we know the composition of a compound, we can usually write its formula. Write:

2 atoms of iron (Fe)

3 atoms of oxygen (O)

Ask for someone to write that as a formula on the board.

The formula is Fe_2O_3

After the formula is written, identify this as the formula for ordinary rust. Its chemical name is iron oxide.

INSTRUCTIONAL PROCEDURE: STUDENT TEAM PROBLEM-SOLVING

Teacher Directions

1. Place students in groups of three or four, including one student who does well in science and one who does not.

2. Direct students to work together in groups to write the formulas and composition of compounds listed on the worksheet.

3. State that the object is to identify the teams who can write chemical formulas and their component parts.

Name _____   **Date** _____

Fun with Formulas

Directions: Use the information given to write the chemical formula.

1. A molecule of water consists of two atoms of hydrogen (H), and one atom of oxygen (O). Write the formula for water.

   _____

2. A molecule of rust, iron oxide, consists of two atoms of iron (Fe) and three atoms of oxygen (O). Write the formula for iron oxide.

   _____

3. A molecule of ordinary table sugar consists of 12 atoms of carbon (C) 22 atoms of hydrogen (H) and 11 atoms of oxygen (O). Write the formula for table sugar.

   _____

4. We have hydrochloric acid in our stomachs to help us digest our food. One molecule of hydrochloric acid consists of one atom of hydrogen (H) and one atom of chlorine (Cl). Write the formula for hydrochloric acid.

   _____

5. Baking soda consists of one atom of sodium (Na), one atom of hydrogen (H), one atom of carbon (C), and three atoms of oxygen (O). Write the formula for baking soda.

   _____

Directions: Use the formula to write the number of atoms and name of each element in the formula.

6. We breathe out the gas carbon dioxide. Its chemical formula is CO_2. There are:

   _____ atom of _____

   _____ atoms of _____

7. Car batteries contain sulfuric acid. Its chemical formula is H_2SO_4. There are:

   _____ atoms of _____

   _____ atom of _____

   _____ atoms of _____

8. The chemical formula for vinegar is CH_3CO_2H. There are:

   _____ atoms of _____

   _____ atoms of _____

   _____ atoms of _____

9. The chemical formula for carbon monoxide is CO. There are:

   _____ atom of _____

   _____ atom of _____

10. The chemical formula for rust is Fe_2O_3. There are:

    _____ atoms of _____

    _____ atoms of _____

4. Distribute Student Worksheet 4-4, "Fun with Formulas." Read the directions together.

5. Direct each team to work on its own for the class period.

6. Review the answers together.

7. Collect and correct the papers.

8. Return the corrected papers and direct the students to keep these papers with the rest of their notes on chemistry.

"If You Like" Assignments

1. Write as many formulas as you can find.

2. Make up your own "Fun with Formulas" problems.

Answers to: Student Worksheet, "Fun with Formulas"

1. H_2O
2. Fe_2O_3
3. $C_{12}H_{22}O_{11}$
4. HCl
5. $NaHCO_3$

6. 1 atom of carbon
   2 atoms of oxygen
7. 2 atoms of hydrogen
   1 atom of sulfur
   4 atoms of oxygen

8. 2 atoms of carbon
   4 atoms of hydrogen
   2 atoms of oxygen
9. 1 atom of carbon
   1 atom of oxygen
10. 2 atoms of iron
    3 atoms of oxygen

Lesson 11. Making and Separating Mixtures

OBJECTIVES

Students will:

1. Define a mixture as a combination of substances that do not combine chemically and can be separated without a chemical change.

2. Make samples of mixtures.

3. Explore ways of separating mixtures.

MATERIALS

Student Worksheet 4-5, "Mixing and Separating Mixtures"

Name _____   Date _____

Mixing and Separating Mixtures

DEFINITION: Mixtures are made up of substances which do not combine chemically and can be separated without a chemical change.

DIRECTIONS: Make the mixtures listed below. Describe how the mixture could be separated without using a chemical change. If you have time, separate your mixture. Use clean pebbles, salt and water for each mixture.

Mixing and Separating Mixtures

Mix	Observations	How They Can Be Separated
Pebbles and salt		
Pebbles and water		
Salt and water		
Pebbles, salt, and water		

On each supply tray:

Large jar of water

Large jar of pebbles or marbles

15 ml of salt

Stirrer

Paper towel

Four empty jars

Tweezers (optional)

Small screen or plastic grid (optional)

Saucer or jar lid

INSTRUCTIONAL PROCEDURES

Lab

Distribute Student Worksheet 4-5, "Mixing and Separating Mixtures." Review the directions. Distribute the supplies.

After the lab is completed, discuss observations and possible ways of separating each mixture.

Lesson 12. The Solution, a Special Kind of Mixture[1]

OBJECTIVES:

The students will define a solution as a special kind of mixture which:

• mixes evenly

• is transparent

• will not separate when filtered

MATERIALS

A teacher-made poster containing the three properties of a solution

A few teaspoons of sugar

A few teaspoons of flour

A bottle of food coloring

Large jar of water

---

[1] Walt Disney Educational Media Company, *Chemistry Matters*, 1986.

Three baby food jars
Stirrer
Three coffee or lab filters

INSTRUCTIONAL PROCEDURES

Review

Say, "We have been studying mixtures. Can someone tell us what a mixture is?" (A combination of two or more substances that do not combine chemically and can be separated without a chemical change.)

Demonstration One/Sugar Water Solution

1. Say, "Today we are going to investigate a special kind of a mixture called a solution."
2. Pour some water in a jar. Add some sugar and stir. Ask, "What happened to the sugar?" (It disappears.)
3. Ask for predictions, "Can we separate the sugar from the water by pouring the mixture through a filter?"
4. Pour the mixture of sugar water through the filter. Invite a lab assistant to look in the filter to see if any sugar has been trapped by the filter; run a finger around inside the filter. Ask, "Is the sugar inside the filter?"
5. (No) "Where is it?" (In the water)

Explanation

The sugar and water:

- mix so evenly that the sugar can no longer be seen
- form a clear or transparent mixture
- do not separate when poured through the filter

This type of mixture is called a solution.

Teacher Directions

Point to the poster listing the three properties of a solution. Review the properties with the class. Direct the students to refer to the poster when trying to identify a substance as one which will form a solution or not. Leave the poster in a prominent place during the labs about solutions.

Demonstration Two/Flour and Water Mixture

Mix flour and water in a jar. Direct the students to use the properties of a solution to decide if flour and water form a solution.

1. Observe it.

   It does not mix evenly.

   It is not transparent; it is cloudy.
2. Pour it through a filter.

   It separates.

Conclusion: flour and water form a mixture, not a solution.

Demonstration Three/Food Coloring and Water Solution

Mix food coloring and water.

1. They mix evenly.
2. The blue water is transparent. We can see through it.
3. They do not separate when filtered. However, some of the food coloring may dye the filter.

Conclusion: food coloring and water form a solution.

Direct the students to copy the properties of a solution from the poster, keep it in their notes, and use it for reference during the next lab on solutions.

Teaching Tip

This lesson may take two days, especially if many students are slow at note taking.

Lesson 13. Identifying and Classifying Items That Will and Will Not Form a Solution with Water[2]

OBJECTIVES

Students will identify and classify items that will and will not form a solution with water.

---

[2] *Ibid.*

MATERIALS

Student Worksheet 4-6, "Identifying and Classifying Items That Will and Will Not Form a Solution with Water"

On each tray:

A large jar of clean water

An empty jar for used water

Clear jars for testing items

Stirrer

One coffee filter for each test

Suggested items to test:

Will form a solution with water	Will not form a solution with water
•sugar cube	•pepper
•powdered soda	•powdered milk
•brown sugar	•corn starch
•vinegar	•cooking oil

INSTRUCTIONAL PROCEDURES

Lab

Ask, "How can we determine which of these items will form a solution with water and which will not?" (Put them in water to see if they:

- mix evenly
- are transparent
- will not separate when filtered)

Tell students to:

1. Use a clean filter, jar, and water for each test.
2. Put used water in the large, empty jar.
3. Record conclusions on the worksheet.

Teaching Tips

Repeat this lab using different items to determine if they will form a solution with water or not.

Explain that there are different solvents, mediums for dissolving, in addition to water. Examples: fingernail polish remover dissolves fingernail polish and turpentine dissolves some paints.

Discuss solutions and mixtures that are foods.

Name _____   **Date** _____

Identifying and Classifying Items
That Will and Will Not
Form a Solution with Water

Forms a solution with water	Does not form a solution with water

"If You Like" Assignments

Find pictures of solutions and mixtures, write identifying labels on them, and display them as a poster, bulletin board, mobile, or collage.

Make a list of solutions and mixtures that are used in the home.

Teaching Tips for Surviving the Last Two Weeks Before Winter/Christmas Vacation

Conduct highly motivating lessons that demonstrate how to make magnificent mixtures for munching and provide students with exciting reasons for coming to school. Once students have been introduced to physical and chemical changes and mixtures, they are ready to learn how to make candy mixtures for munching and giving.

Before Teaching Candy Making

1. Decide who will pay for the supplies. Consult your principal and district policy to determine if it will be each student, the building budget, the PTA, or some type of district grant.
2. If the students are responsible for their own supplies, make some arrangements for those who cannot afford to pay for the supplies.
3. Prepare letters of explanation and invitation to parents and students. Clear the letter with your principal. A sample letter is provided at the end of this section, on page 132.
4. Reproduce copies of the recipes that you select. Sample recipes are provided at the end of this section, pages 133 through 136.
5. Purchase the supplies that you will need.

Lesson 14. Introducing Make-and-Take Workshops for Making Magnificent Mixtures for Munching and Giving

OBJECTIVES

Students will:

1. Read and discuss the letter to students and parents explaining the regulations and directions for "Make-and-Take Workshops."
2. Read and discuss the recipe for Butter Cream Mixture.
3. Identify making butter cream as an example of a physical change and a mixture.
4. Identify eating butter cream as an example of a chemical change.

MATERIALS

Letters to students and parents explaining the "Make-and-Take Workshops." See page 132.

One copy per student of the recipe for Butter Cream Mixture. See page 133.

Two three-foot pieces of waxed paper on a tray

One block (¼ pound) of butter or margarine, room temperature

One 16-ounce box of confectioners sugar

One ¼ cup of table sugar in a plastic bag

One small container of chocolate jimmies

One small bag of dried apricots

INSTRUCTIONAL PROCEDURES

Introduce the Make-and-Take Workshop

Explain that you are going to:

1. Make some magnificent mixtures that are great for munching and gift giving.

2. Invite everyone to have a small taste to discover who likes which mixtures.

3. Provide written recipes so they can be made in class and at home.

4. Invite everyone to make one or two recipes during the last two science classes before the winter vacation; take them home, enjoy some, and consider giving some as gifts.

5. Invite those who do not want to make a mixture to become lab assistants.

6. Require everyone to explain how to make the recipes and identify each as an example of a physical change.

Demonstration / Making Butter Cream

Getting ready: make sure that you have nothing on your hands. For example, no rings or fingernail polish. Explain to the students that if they decide to make this mixture, they also will have to come to class with no rings or polish. Wash your hands. Keep the paper towel that you used to dry your hands. It will be useful after you have finished.

Making a Butter Cream Mixture

1. Place the ¼ pound of butter or margarine on the tray covered with waxed paper. Have students make observations. (It is yellow. Soft to the touch.)

2. Pour ½ of the box of confectioners sugar on the butter. Have students make observations. (It is a white powder.)

3. Mix the butter and the sugar with your fingers. Add some of the remaining sugar. You will use a little less than the contents of the box. Have the students make observations. (It makes pale yellow crumbs.)

It Needs to Be Kneaded

4. As it turns pale yellow and begins to mix and make "crumbs," pick some of it up in your hands. First "pack" it as though you were making a snow ball. As it starts to resemble dough, knead it. It needs to be kneaded. When it no longer sticks to your hands, it is done. The butter and confectioners sugar mix to make a butter cream. This takes about five to ten minutes.

5. Place a generous ball of the butter cream in your hand. Take a pinch and taste it. Offer a taste to all students so they can determine if they like it or not.

6. Remind them that they may pass if they do not want to taste it.

Demonstrate How to Present It:

- Roll a pinch into a small ball, shake it in the plastic bag of table sugar.
- Place a pinch in a dried apricot, shake in the table sugar.
- Roll a small ball in chocolate jimmies.

Invite each student to make one of the above. They may eat it or save it to take home and show the family.

While each student makes a butter cream candy, have two students distribute the recipes and the letters to students and parents. List the ingredients and their costs on the board. Direct students to use the costs listed on the board to calculate the cost of one batch of this candy.

Demonstrate How to Clean Up:

- Roll up waxed paper and throw it away.
- Place butter cream candies in a nonbreakable container.
- Clean the work space.
- Wash hands.

Read and discuss the letter. Answer any questions about the directions for the Make-and-Take Workshop.

Read and discuss the recipe for Butter Cream. Explain that for the recipe, they will need one-half of a ¼ pound *block* of butter and one-half box or 6 ounces of powdered sugar. Discuss how to halve or double the recipe.

Explain that if the butter cream looks and feels like a white brick, there is too much sugar. Merely add some butter. If it doesn't come off your hands, add some sugar.

Lesson 15. How to Make Peanut Butter Candy

OBJECTIVES

Students will:

1. Explain how to make peanut butter candy.
2. Identify making peanut butter candy as an example of a physical change.
3. Identify eating peanut butter candy as a chemical change.
4. Calculate the cost of one batch of peanut butter candy.
5. Calculate halving and doubling the recipe.

MATERIALS

Tray covered with waxed paper

⅔ cup of peanut butter

One cup of powdered skim milk

⅔ cup of confectioners 10 - X powdered sugar

Two tablespoons of vanilla extract

Two to four tablespoons of water

Measuring spoons

Recipes for Peanut Butter Candy (page 134)

INSTRUCTIONAL PROCEDURE

Demonstration

Measure each ingredient as students watch. Place the peanut butter on the tray. Add powdered milk, sugar, vanilla and water. Mix with fingers. Ask for observations. (A mess. The brown and white colors blend to make a creamy brown. It sticks to the hands.)

After mixing with your fingers, pick it up with your hands and knead. When the mixture is a creamy brown and no longer sticks to your hands, it is done. If it feels dry, not creamy, it needs one or two more tablespoons of water.

Hold a handful. Take a small pinch and taste it. Offer each student a small taste.

Demonstrate Different Ways to Present It:

- Roll it into small balls.
- Shake a ball in a plastic bag of table sugar.
- Roll a ball in chocolate jimmies.

Distribute copies of the recipe. List the prices of individual ingredients on the board. Have students calculate the cost of making a batch while two or three students make one piece of candy.

Review

Making peanut butter candy is an example of a physical change and a mixture. Eating it is an example of a chemical change.

Teaching Tips for Workshop Days

1. Provide the vanilla extract for each batch of peanut butter candy rather than have the students bring a small amount of vanilla to school.
2. The teacher or another adult adds two or three tablespoons of water to each batch of candy. After the students work the mixture, the teacher feels the mixture to determine if it is creamy. If it is not creamy, add one or two more tablespoons of water.

Lesson 16. How to Make Chocolate Peanut Clusters

OBJECTIVES

Students will:

1. Identify melting chocolate as an example of a physical change.
2. Identify burning chocolate as an example of a chemical change.
3. State different ways of changing this recipe.
4. Calculate the cost of making clusters.

MATERIALS

Hot plate

Saucepan

Potholder

Cooking spoon

Two large cookie sheets covered with waxed paper

Twelve ounces of semi-sweet chocolate morsels or sweetened carob

Twelve ounces of salted, Spanish peanuts, or any nuts or raisins

Recipes for Chocolate Covered Peanut Clusters (page 135)

INSTRUCTIONAL PROCEDURES

Demonstration

1. Place twelve ounces of chocolate in a saucepan. Have students make observations. (Good is not an acceptable observation. Rather, they might say Brown. Pieces.) Ask, "What type of change occurs when the chocolate melts?" (Physical)

Ask, "What type of change would it be if I were to accidentally burn the chocolate?" (Chemical. The chocolate would turn chemically to carbon, water vapor, and gases.)

2. Stir the chocolate while melting it. It takes about five minutes to melt the chocolate. Have the students make observations. (Creamy looking. Brown liquid.)

3. Fold in twelve ounces of Spanish peanuts. Walk about the class as you do, explaining the term "fold" and giving students a close look. Have them make observations. (The peanuts are covered by the liquid chocolate.) This will take about five minutes.

4. Spoon a few tablespoon-size measurements of the peanut and chocolate mixture on the waxed paper. Have the students make observations. (Shiny, wet globs of chocolate and peanuts.)

5. Explain that they will be ready to pick up and eat when the melted chocolate cools and returns to a solid. The clusters will look dull, not shiny. It takes about two or three hours.

6. Distribute the copies of the recipes. List the prices of the individual ingredients on the board. Have the students calculate the cost while each student makes one peanut cluster.

REVIEW AND DISCUSS

Review how to halve the recipe and directions for the workshop days. Discuss substances that could be substituted for the semi-sweet chocolate: milk chocolate, sweetened carob, butterscotch morsels. Discuss substitutes for the Spanish peanuts: any type of nut, raisins, small marshmallows, small pretzels, and some cereals. Explain that the recipes can be changed because these candies are all examples of physical changes: melting, mixing, solidifying.

Teaching Tips

1. The first school year that you teach candy making, consider teaching only one or two recipes.
2. Collect your own recipes stressing nutritious snacks.
3. Invite another teacher and class to make candy with you.
4. Send a few samples to the office and custodial staff.

5. Invite parents to candy demonstration classes and workshop days.
6. Have an adult control the hot plate on the workshop days.
7. Introduce workshop days about two weeks ahead of time to allow parents and students time to plan.
8. Alternate candy making demonstrations with labs about mixtures and reviews of physical and chemical changes.

Lesson 17. The Day Before the Make-and-Take Workshops

OBJECTIVES

Students will:

1. State the workshop directions.
2. Explain how to make the candies that have been demonstrated.
3. Explain how to set up and clean up for candy making.
4. Explain that all ingredients need to be measured at home and packed in nonbreakable containers.
5. State that hands must be clean and free of nail polish and rings.

MATERIALS

Copies of recipes
Letters to parents and students

INSTRUCTIONAL PROCEDURES

Review all directions for the workshop; such as:

• Measure and pack all ingredients at home on the night before the workshop.
• Bring six to twelve feet of waxed paper for work and easy cleanup.
• Use nonbreakable containers.
• Print name and homeroom number on the package.
• No rings or nail polish.
• Long hair tied back.
• Deliver all packages to the classroom the morning of the workshop.
• Package butter cream and peanut butter candy as soon as it is finished and clean up the work areas.

"If You Like" Assignments

1. Make these and other candies or cookies at home, with parents' permission, during the winter break.

2. Write your own cookbook.

3. Write a list of changes and identify each as a physical or chemical change.

Lesson 18. Managing the Workshop Day

OBJECTIVES

Students will:

1. Follow written and oral workshop directions.

2. Make or help someone make a candy mixture.

3. Clean up and take home whatever they made.

MATERIALS

Waxed paper

Aluminum foil

Large vanilla extract

Measuring spoons

Hot plate with an adult in charge of it

A cup of water

A small amount of each ingredient to supplement anyone's needs

Paper towels

Soap

INSTRUCTIONAL PROCEDURES/HANDS-ON WORKSHOP

Directions

1. Introduce the adult who will operate the hot tray.

2. Identify those who will be lab assistants.

3. Tell half of the students to set up their work area by putting out waxed paper and then setting out all of the ingredients. Tell the others to wash their hands. Switch jobs.

4. Tell those who are making mixtures that do not require heat, to begin once their hands are clean and their candy making area is set up.

5. Direct those who need the hot plate to line up, two at a time, at the hot plate, with the chocolate in their own pan. The adult melts the chocolate. The student returns the pan with the melted chocolate to the work area and folds in the other ingredients.

Teaching Tips

1. Once the workshop begins, test each student's candy by kneading it. Once you have made the candy, you will know by feeling it if it is the right consistency or not.
2. If the butter cream is dry, add butter. If it won't come off your hands, have an assistant add a small amount of powdered sugar.
3. If the peanut butter candy is dry, add a tablespoon or two of water.
4. Wear comfortable shoes and wash-and-wear clothing.

Lesson 19. Making Winter Weather Bird Feeders

OBJECTIVES

Students will:

1. Explain how to make nutritious mixtures for birds.
2. Design at least one holder for the mixture.
3. Identify the making of winter weather bird feeders as examples of physical changes.

MATERIALS

Large clear glass bowl

Waxed paper for a work surface

Large spoon

One coffee tin of chilled meat drippings

Small pieces of suet (optional)

One cup of peanut butter

One to three cups of bird seed

One cup of powdered milk

One cup of stale bread, cookies, or crackers (optional)

Many feet of string

A container for the mixture: large pine cones, empty grapefruit halves, or milk cartons

INSTRUCTIONAL PROCEDURES

Review

Ask for someone to review the definition of a mixture. (A mixture is a blend of two or more substances that don't combine chemically, and which can be separated without a chemical change.)

Explain that the measurements need not be exact when making a mixture, but that they must be exact when making a compound.
Ask:

1. Where do birds get their food in the winter after the ground freezes? After it is covered with snow?

2. Ask who feeds birds or other wild animals on a regular basis in the winter?

Explain that you are going to demonstrate how to make a food mixture that is good for birds, squirrels, and other small wild animals. Cats and dogs frequently enjoy these winter dishes so they have to be placed out of reach of pets. For safety reasons an adult should hang the feeder in a tree or on a pole.

Demonstration/The Mixture

1. Cover the work area with waxed paper.

2. Place a clear, glass bowl so that all can see it.

3. Identify each ingredient, tell how you obtained it as you put it in the bowl.

4. Ask for observations.

5. Mix all of the ingredients.

6. If it is too creamy or runs, add more powders to help hold it together. When it is the consistency of chunky peanut butter, it is ready to put in a bird feeder.

Demonstration/Making the Feeders

1. Place a string on a pine cone so that the pine cone can be hung from a tree.

2. Spread the mixture into and around the crevices on the pine cone until it can hold no more.

3. Roll it in bird seed.

4. Wrap it in waxed paper and put it in a refrigerator until an adult can safely hang it in a tree or on a pole.

POINTS FOR DISCUSSION

1. The measurements do not have to be exact to make a nutritious meal because this is a mixture, not a compound.

2. Making it is an example of a physical change.

3. Eating is an example of a chemical change. When the birds eat this mixture, their digestive system chemically turns it to proteins, fats, and carbohydrates which are used for growth and energy.

Safety Concerns

A fall from a tree or other high place can cause serious injury. Students need to be instructed to have parents hang the feeders.

Teaching Tip

Reproduce these recipes. If you have access to a computer and printer, enter the recipe for winter weather bird feeders on a floppy disk. Include safety directions for hanging the feeder. Permit interested students to print out a copy of the directions to take home. This provides excellent motivation for some students. This is an excellent lesson with which to begin the New Year.

"If You Like" Assignments

- Make winter weather feeders throughout the remainder of the winter.
- Keep a photographic and/or written record of the animals who visit your feeder.
- Prepare a talk about the animals who visit the feeder, the types of food mixtures and/or types of feeders.

Lesson 20. Making Carbon Dioxide to Review Writing a Chemical Equation and Chemical Changes

OBJECTIVES

Students will:

1. State that vinegar and baking soda combine chemically to make carbon dioxide.
2. State that baking soda is excellent for extinguishing fires, especially grease fires.
3. Write the balanced equation for making carbon dioxide using vinegar and baking soda.

MATERIALS

Cafeteria tray
Two candles each melted to a lid
Matches
A box of baking soda
A bottle of vinegar

Two flasks

A balloon

INSTRUCTIONAL PROCEDURES

Demonstration One/Vinegar + Baking Soda Makes Carbon Dioxide

Place three tablespoons of baking soda in a flask. Add two tablespoons of vinegar. Place a balloon over the bottle quickly. Shake the bottle. The balloon begins to fill. Ask for observations. (The vinegar and baking soda fizz. The balloon fills.)

Ask for inferences. (A gas is made which fills the balloon.)

Explanation

The vinegar and baking soda chemically combine to produce the colorless, odorless gas. As the gas is released, it enters the balloon.

Demonstration Two/Carbon Dioxide Extinguishes a Flame

1. Using standard safety practices, light the two candles. Hold one candle near the mouth of the flask. *Observation.* (The candle remains lighted.)

2. Pour three tablespoons of baking soda in the flask. Hold the candle near the opening. *Observation.* (The candle remains lighted.)

Hold the candle near the open bottle of vinegar. *Observation.* (The candle remains lighted.)

Pour a few tablespoons of vinegar in the flask. It fizzes. Hold the candle near the opening but do not let the candle get wet. *Observation.* (The flame goes out.)

Inference. (Whatever blew up the balloon also put out the flame.)

Explanation

Vinegar and baking soda chemically reacted to produce carbon dioxide which does not support a flame. When the carbon dioxide surrounds the flame it prevents oxygen from getting to the flame. Thus, carbon dioxide smothers a flame. We cannot see carbon dioxide but we can see its effects on the balloon and the flame.

Review Writing Like a Chemist (Grades 5–8)

"What form would a chemist use to write this reaction?" (An equation.)

Use the four steps to write this equation. While you write on the board, direct the students to write the four steps on paper.

ON THE BOARD

1. Key words:

    baking soda + vinegar → carbon dioxide + sodium acetate + water

2. Write the formulas:

    $NaHCO_3 + CH_3CO_2H \rightarrow CO_2 + NaOOCCH_3 + H_2O$

3. Is it balanced? (Yes)

1 Na	1 Na
5 H	5 H
3 C	3 C
5 O	5 O

4. The balanced equation is:

    $NaHCO_3 + CH_3CO_2H \rightarrow CO_2 + NaOOCCH_3 + H_2O$

Teaching Tips

1. Have everyone write the four steps but do not require anyone to memorize the balanced equation. Let it be a challenge for those who are interested.

2. Have an "artist" make a poster of the four steps for balancing this equation.

3. Discuss:

    A. Carbon dioxide is released when a bottle of soft drink is shaken or opened.

    B. Carbon dioxide is released when baking soda is put on a fire. Thus, it is a good idea to apply this knowledge of chemistry and keep baking soda in the kitchen for small cooking and grease fires.

Lesson 21. The Chem Mystery of Floating and Sinking Moth Balls

OBJECTIVES (Grades 4–8)

Students will use information obtained from observations to solve the chem mystery of the continuously floating and sinking moth balls.

MATERIALS

Copies of Student Worksheet 4-7, "The Chem Mystery of the Floating and Sinking Moth Balls"

On each tray:

    A jar of water

    One tablespoonful of vinegar in a container

Name _____   Date _____

The Chem Mystery of the Floating and Sinking Moth Balls

Directions: Add one tablespoon of vinegar and one tablespoon of baking soda to a glass of water. Add three moth balls. Record your observations. Write at least one inference that would explain the movements of the moth balls.

Moth Balls in a Mixture of Water, Baking Soda, and Vinegar

Observations	Inference

One tablespoon of baking soda in a container

Three moth balls

INSTRUCTIONAL PROCEDURES

Lab Directions

Distribute the Student Worksheet, 4-7, "The Chem Mystery of Sinking and Floating Moth Balls." Read and discuss the written directions.

1. Add the vinegar and the baking soda to the jar of water.
2. Place the moth balls in the liquid in the jar.
3. Observe their movements.
4. Work in lab teams and write a hypothesis that would explain the mysterious movements of the moth balls.

Explanation

The vinegar and baking soda combine chemically to make bubbles of carbon dioxide. The bubbles collect on a moth ball causing it to float to the top. While it is floating at the top, some of the carbon dioxide bubbles burst into the air causing the moth ball to lose some of its buoyant bubbles and sink. The vinegar and baking soda produce more carbon dioxide bubbles which gather on the moth ball causing it to float to the top again where it will eventually lose its bubbles and sink. This will continue until the vinegar and baking soda stop producing carbon dioxide.

Lesson 22. Green Plants, Nature's Chemical Factories

OBJECTIVES

Students will:

1. Take notes, including a sketch, explaining the process of photosynthesis.
2. State that green plants make their own food.
3. Identify oxygen as a chemical product of green plants that is essential for life.
4. Identify at least two contributions that green plants make to animal life.

MATERIALS

Board and chalk

Paper and pencil

INSTRUCTIONAL PROCEDURES

Look, Listen, and Take Notes

Explain that the green plants are nature's chemical factories. That is, they use energy from sunlight to chemically make their own food.

In their green leaves, plants chemically separate water and carbon dioxide into component parts and recombine them into food and other products. The scientific term for a green plant chemically making its own food is photosynthesis. "Photo" means light; "synthesis" means to make. Thus, photosynthesis means to make with light. Fortunately for us, when the green plant makes its own food, it makes food and oxygen that we need.

Sketch and Explain

Next, sketch an apple tree on the board. Direct students to copy it and anything else that you write on the board. Ask, "What does the plant take in through its roots?" (Water) Write "water" near the roots. See Figure 4-4, Nature's Chemical Plant.

Ask, "What does the leaf take in through the openings on its underside?" (Carbon dioxide) Add carbon dioxide near the underside of a leaf. See Figure 4-4.

Direct students to:

1. Write the chemical formulas under the names on their drawings.
2. Check the formulas with those that you have written on the board.
3. Make any corrections if necessary.

Next, explain that many chemical changes go on in the green leaf. The leaf gets its green color from a pigment named **chlorophyll**. This green dye is very interesting because it is needed for chemical changes to take place, but the chlorophyll itself does not change chemically. Chemists call this type of changing agent a **catalyst**.

In nature's green factory, water is separated into hydrogen and oxygen. Carbon dioxide is separated into carbon and oxygen. Through a series of chemical changes, the plant combines the carbon, hydrogen, and oxygen to make new products. For us, the two most important products are oxygen and a form of sugar named glucose. Glucose is the sugar found in fruits and honey, not white table sugar.

On the sketch, near the underside of a leaf, write the word and the chemical symbol for oxygen. See Figure 4-4.

Sketch a few seeds, and an apple formed around the seeds. Write the word "glucose" and its chemical formula, $C_6H_{12}O_6$ near the apple.

Stress the importance of photosynthesis to us: green plants supply us with a continuous supply of oxygen and food.

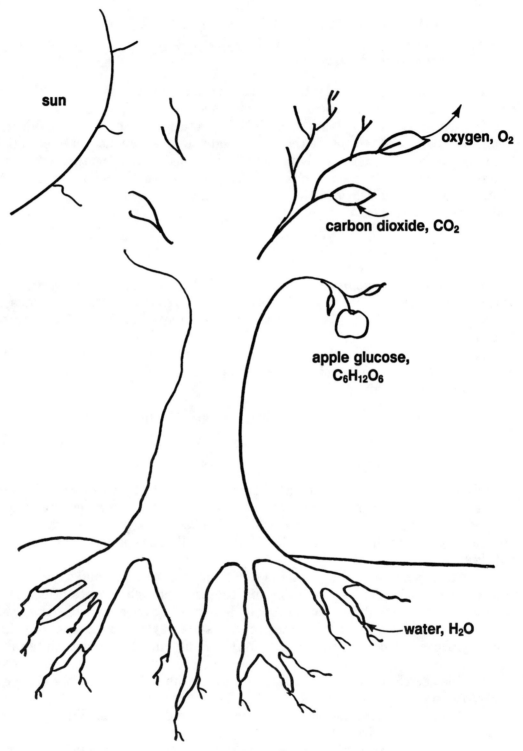

sun

oxygen, O_2

carbon dioxide, CO_2

apple glucose, $C_6H_{12}O_6$

water, H_2O

Figure 4-4. Nature's chemical plant

126

Lesson 23. Writing the Equation for Photosynthesis

OBJECTIVES

Students will:

1. Explain the process of photosynthesis using a sketch of a fruit tree.
2. Write the balanced equation for the process of photosynthesis using the four steps to writing an equation.

MATERIALS

A sketch with notes introducing the process of photosynthesis.
A new sketch of a fruit tree on the board.

ON THE BOARD

1. Key words:
   water plus carbon dioxide chemically makes glucose and oxygen
2. Formulas:
   $$H_2O + CO_2 \rightarrow C_6H_{12}O_6 + O_2\uparrow$$
3. Is it balanced? (No)

2 H	12 H
3 O	8 O
1 C	6 C

4. The balanced equation is:
   $$6 H_2O + 6 CO_2 \rightarrow C_6H_{12}O_6 + 6 O_2\uparrow$$

INSTRUCTIONAL PROCEDURES

Review

Ask questions and write the answers at appropriate places on the new sketch of an apple tree on the board:

1. What does a green plant take in through its roots? (Water)
2. What does it take in through its leaves? (Carbon dioxide)
3. What types of changes take place within the leaf? (Chemical)
4. What type of energy does the plant need for the chemical changes? (Light)
5. Name one of the end products of photosynthesis. (Oxygen)
6. What is the chemical symbol for oxygen? (O)

7. Name the other product that we studied. (Fruit sugar or glucose)

8. What is the formula for glucose? ($C_6H_{12}O_6$)

Explain that chemists express this process with a series of equations. The class is going to write the equation most frequently written to express the process of photosynthesis.

Write the four steps for writing an equation on the board and direct the students to copy the four steps in their notes. Begin by asking, "What are the key words for explaining photosynthesis?" Write them on the board while students write them on their paper.

Ask, "What is the formula for water?" Write it under the word, "water."

Continue until the four formulas are written. Balance the equation for the students.

Teaching Tips

1. Have students make posters illustrating the processes of photosynthesis, complete with the equation.

2. Frequently, equations are written with no space between the formulas and the numbers necessary to balance the equation. Since the number is not a part of the formula, it is recommended that you start your students with the good habit of putting a space between the numbers and the symbols.

3. Balance the equation for the students. (Do not require students to balance this equation. Balance it for them.)

4. Discuss the importance of photosynthesis to all living creatures.

5. Discuss the importance of protecting green plants and their environment.

For the Teacher

The end products of photosynthesis are glucose, oxygen, and water. Water is usually not included in the equation until junior high or high school science classes because the equation becomes more complex. It is provided here for your information.

$$6\ CO_2\ +\ 12\ H_2O\ \rightarrow\ C_6H_{12}O_6\ +\ 6\ O_2\ +\ 6\ H_2O$$

Lesson 24. Why Do Leaves Turn Color in the Fall?

OBJECTIVES

Students will:

1. Observe yellow and orange pigments that are present in a green leaf or a green dye.

2. State that green leaves reveal other coloring agents when chlorophyll is no longer present.

3. State that green leaves slow, and eventually stop, their production of chlorophyll as the length of daylight decreases and the temperatures cool.

MATERIALS

A variety of green leaves and for each leaf:

One jar with an inch of rubbing alcohol (Caution: toxic)

One strip of coffee filter paper

One ruler

See Figure 4-5, Paper Chromatography.

INSTRUCTIONAL PROCEDURES

Demonstration with Green Leaves

For each green leaf, make the following setup:

1. Scratch the leaf. Rub some of the green coloring onto the filter paper, about an inch from the bottom of the paper.
2. Wrap the strip of paper around a ruler.
3. Lower the paper into the alcohol and rest the ruler on the sides of the jar.
4. The rubbing alcohol should not touch the leaf rubbing.
5. Observe every 20 to 30 minutes.

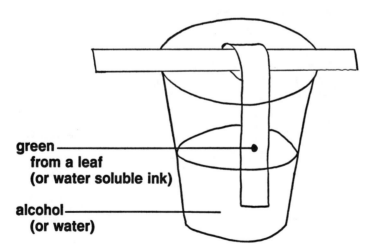

green
    from a leaf
    (or water soluble ink)

alcohol
    (or water)

Figure 4-5. Paper chromatography

Explanation

The filter paper will slowly absorb some of the alcohol. The alcohol will dissolve the colorings in the leaf rubbing. The dissolved colorings will separate out, dying the filter paper, and showing the other colors that were present but hidden by the green chlorophyll.

Safety Concern

Since rubbing alcohol is toxic when swallowed, this is recommended only as a teacher demonstration.

Optional Demonstration or Lab with Water Soluble Ink

If you teach this lesson during a time of year when green leaves are scarce, or you want a hands-on activity, use a different type of paper chromatography.

MATERIALS

Water soluble, green felt tip pens
Water in a jar
Ruler
Filter paper or strips of paper towels

Explanation

The water dissolves the green coloring. The different colors that are present will separate out and dye the filter. When the colors were mixed, the green was the only color that showed. These colorings behave similarly to those in a green leaf.

Teaching Tip

Experiment with different brands of water soluble, green markers to identify brands that have yellow and other dyes mixed with the green dye.

Lesson 25. Review Physical and Chemical Changes with Food Coloring, Water, and Bleach

OBJECTIVES

Students will:

1. Identify adding food coloring to water as an example of a physical change.

2. Identify adding bleach to a mixture of food coloring and water as an example of a chemical change.

MATERIALS

Two jars
Food coloring
A few ounces of bleach
Jar of water

INSTRUCTIONAL PROCEDURES

Demonstration

Pour some water in a clear jar. Add a few drops of food coloring. Ask, "What kind of change is this?" (Physical)

Add some bleach to this mixture. The coloring disappears. Ask, "What kind of change is this?" (Chemical)

Review examples of physical and chemical changes by repeating some of the demonstrations and by asking such questions as, "What kind of change occurs when ..."

1. Paper is wet? (Physical)
2. Paper is burned? (Chemical)
3. Wax is melted? (Physical)
4. Ice is melted? (Physical)
5. Chocolate is melted? (Physical)
6. Chocolate is burned? (Chemical)
7. Baking soda combines with vinegar? (Chemical)
8. Iron rusts? (Chemical)
9. Green plants produce glucose and oxygen? (Chemical)
10. Water turns to steam? (Physical)

For Your Consideration

During the last week of school, consider a review of physical and chemical changes. On the last full day before the summer recess, make an easy, no-cook ice cream in the morning. Discuss the physical changes that take place while making it. Enjoy eating the ice cream during the afternoon and discuss the chemical change caused by digestion. What a sweet way to conclude a school year and an introduction to chemistry.

A recipe is included at the end of this section.

Sample Letter to Parents and Students

December _____

Dear Parents and Students:

During beginning chemistry classes to be held this week, we will investigate making candy as examples of physical and chemical changes. I will demonstrate how to make a few candies and provide written recipes.

Students are invited, not required, to make their own candy in science class on _____:

_____
        (Day)   (Date)   (Time)

and on _____.
           (Day)   (Date)   (Time)

To make candy, students need to:

- gather their own ingredients and supplies
- measure all of the ingredients at home
- pack the measured ingredients along with 6 feet of waxed paper and nonbreakable containers in a bag
- write their name and homeroom number on the bag
- deliver the supplies to the science class on the morning of the candy-making workshop before the first class starts.

Parents are welcome to join us during one or both of these special classes.

We need one or two parents to volunteer to supervise the use of the hot plate.

If you have any questions, please call the school and leave a message, including your phone number. I will return your call as soon as possible. The school phone number is _____.

Sincerely,

Teacher

Principal

- -

Please fill in and return:

_____ I will visit my child's class on _____ at _____.
                                            (Date)   (Time)

_____ I will not visit my child's science class.

_____ I will be willing to supervise the use of the hot plate.

Student's name _____

Parent's signature _____

Recipe for Butter Cream Mixture

Ingredients

4 tablespoons of butter or margarine at room temperature

8–10 ounces of confectioners sugar (10 X powdered)

4 tablespoons of table sugar in a plastic bag

A 3-ounce container of chocolate jimmies (optional)

An 8- to 12-ounce pack of pitted dates or dried apricots

A nonbreakable mixing bowl or waxed paper

Directions

1. Wash your hands.
2. Place 4 ounces of butter on the waxed paper.
3. Pour about half of the confectioners sugar on the butter.
4. Mix with your fingers.
5. Have a friend add some of the confectioners sugar, a little at a time, while you mix the butter and the sugar. (You may not need all of the sugar.)
6. Place the mixture in the palm of your hands and mix.
7. Knead the mixture until it no longer sticks to your hands.
8. After kneading, if it sticks to your hands, it needs a little more confectioners sugar.
9. Take a small pinch of the mixture and roll it into a ball.

Be creative

1. Toss a butter cream ball in the table sugar.
2. Roll a butter cream ball in chocolate jimmies.
3. Place a small pinch of the mixture in a dried apricot or date and toss in table sugar.
4. Neatly arrange your homemade candies on a nonbreakable plate. Cover with plastic wrap.

Recipe for Peanut Butter Candy

Ingredients

⅔ cup of peanut butter
1 cup of powdered skim milk
⅔ cup of confectioners sugar
⅓ cup of butter or margarine
2 tablespoons of vanilla
2 to 3 tablespoons of water
nonbreakable mixing bowl or waxed paper

Directions

1. Wash your hands.
2. Place all of the ingredients on the waxed paper.
3. Mix with your fingers. Knead the mixture until it won't stick to your hands.
4. Roll into small balls and enjoy.

Be creative

1. Roll in table sugar.
2. Roll in chocolate jimmies.

Place on a nonbreakable plate and cover with plastic wrap.

Chocolate Covered Peanut Clusters

Ingredients

    8 ounces of Spanish peanuts
    8 ounces of semi-sweet chocolate
    Pan, spoon and potholder
    Two cookie sheets covered with waxed paper

Directions

1. Wash your hands.
2. Have an adult melt the chocolate. It will take about 5 minutes.
3. Remove the chocolate from the heat.
4. Fold in the Spanish peanuts.
5. Spoon small portions onto the cookie sheets.
6. It will take about 2 hours to dry and harden and appear dull.

Be creative

1. In place of semi-sweet chocolate, try sweetened carob or milk chocolate.
2. In place of the Spanish peanuts, try a different kind of nut, raisins, or bite-sized marshmallows.

No-Cook Vanilla Ice Cream (Philadelphia)[3]

Ingredients

> 1 quart thin cream ("Half & Half")
> 1 tablespoon vanilla
> ¾ cup of sugar
> 3 cups of whole milk
> ⅛ teaspoon of salt

Directions

Thoroughly mix all ingredients. Then freeze.
Makes 1½ quarts.

Be creative

Add one or more of the following:

- Fruit
- Jimmies
- Hot Fudge
- Butterscotch

---

[3] Recipes and Instructions for Homemade Ice Cream: Richmond Cedar Works Mfg. Corp., 400 Bridge Street, Danville, Virginia 24541.

Graphs: Great
Grids for Communicating

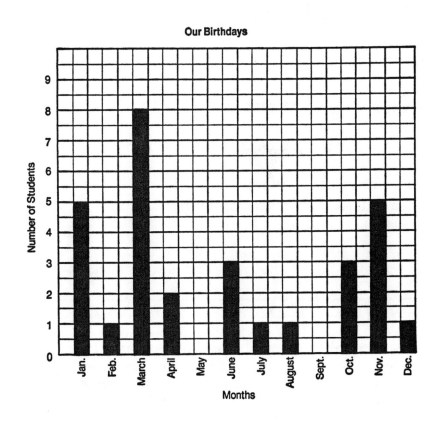

Personal Perspective

Graphs are truly great grids for communicating. Think about the ways that scientists and businesspersons use graphs. They make presentations using tables and graphs to communicate their findings and to make recommendations and predictions. Their presentations can be in the form of oral or written reports.

Teaching students to use graphs to communicate findings in oral and written presentations, and to interpret graphs made by others, provides them with lifetime skills. These skills should begin in the elementary grades and be used throughout and beyond formal schooling. When taught as effective tools for interpreting data, these skills can be helpful in training wise consumers and informed citizens.

Graphing can be taught effectively and learned easily if it is kept out of the realm of the abstract. It should be presented as a part of a realistic investigation. The investigation can be actual or vicarious. That is, the students can actually conduct an investigation in which they gather, graph, and interpret data. Or, you can provide them with a worksheet describing an investigation and providing data which they will graph and/or analyze. A combination of actual and vicarious graphing investigations is recommended.

Five Important Steps for Graphing

To teach graphing as a part of an investigation, the following steps are helpful:

1. Identify the area of investigation.
2. Gather and record information on an organized table.
3. Transfer that data to a graph.
4. Interpret the graph.
5. Present the findings, formally or informally. (Optional)

Is There Graphing Beyond School?

When the first three steps are skipped, students rarely know what to do with the graphs that they make other than turning them in for a grade. Consequently, they view graphs as class assignments and see little use for them outside the classroom. When the five steps are taught, students understand how graphs are used in and out of the classroom. They view class presentations as logical. In the process, they develop an understanding that there is graphing beyond school.

Possible Presentations

If your school participates in a science fair, you then have an excellent forum for students to make formal presentations. If your school does not, students can make formal presentations for P.T.A. Open House, other classes, and their classmates.

Lesson 1. Graphing as Part of an Investigation

OBJECTIVES

Students will copy and complete a bar graph started by the teacher.

MATERIALS

Chart paper for a table of data
Transparency of a grid
Overhead projector

INSTRUCTIONAL PROCEDURES

Demonstration

Step 1. Identify the investigation.

Tell the students that you are going to take a survey of the months in which they have their birthdays and demonstrate how to graph the results. Ask how many students have birthdays in January.

Step 2. Record the data on an organized table.

Record the results on the table. Repeat the question for all of the months and record the information. See Figure 5-1.

Step 3. Transfer the Data to a Graph.

   a. Demonstrate how to graph the data for the months of January, February, March, and April.
   b. Direct students to copy and complete the graph. See Figure 5-2.

A SAMPLE GRAPH

While students are working, check graphing progress for:

1. Labeling both axes.
2. Writing numbers on lines, not in spaces.
3. Writing a title.
4. Writing a zero at the intersection of the two axes.
5. Making the bars correctly.

Our Birthdays

Months	Number of Students
Jan.	5
Feb.	1
Mar.	8
Apr.	2
May	0
June	3
July	1
Aug.	1
Sept.	0
Oct.	3
Nov.	5
Dec.	1

Figure 5-1. Sample table of data

Our Birthdays

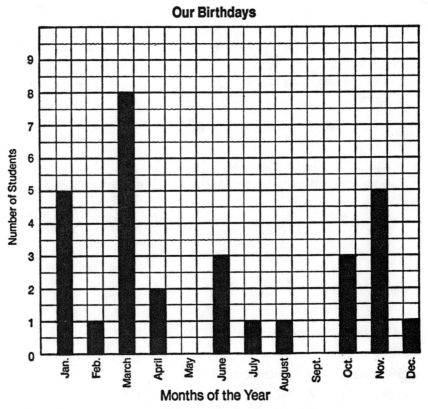

Figure 5-2. Sample graph: Our Birthdays

Step 4. Interpret the data on the graph.

Look at the graphs and discuss the findings. Ask:

1. Do we have any months in which no one has a birthday?
2. In which month do we have the most birthdays?
3. How many students don't have to go to school on their birthdays because their birthdays are in the summer or on a holiday?

Step 5. Present the findings.

Direct some students to put the graphs along with the table of data on a bulletin board.

Teaching Time

For young students this lesson may require two days.

Lesson 2. Graphing the Results of a Survey

OBJECTIVES

Students will:

1. Construct a bar graph.
2. Transfer the data from a table of data to a bar graph.

MATERIALS

Overhead projector
Grid transparency
Graph paper
Rulers

ON THE BOARD

Table of Data. See Figure 5-3.

What Is Your Favorite Flavor of Ice Cream?

Student	Favorite Flavor
1	Chocolate
2	"
3	"
4	"
5	Vanilla
6	Chocolate
7	"
8	Strawberry
9	Chocolate
10	"
11	"
12	"
13	"
14	"
15	Vanilla Fudge
16	Banana
17	Chocolate
18	"
19	"
20	Strawberry
21	Chocolate
22	Vanilla
23	"
24	Strawberry
25	Chocolate
26	Vanilla
27	Chocolate
28	Vanilla
29	Chocolate
30	Vanilla

Figure 5-3. Sample table of data

INSTRUCTIONAL PROCEDURES

1. Identify the investigation as a survey to determine the favorite ice cream flavor of the students in your class. Ask, "What is your favorite flavor of ice cream?"

2. Record each person's response on the table of data.

3. Graph the results of the survey:

    a. Demonstrate how to construct and label the axes.

    b. Demonstrate how to use the three or four most common names of ice cream and identify the rest as "Other." See Figure 5-4.

Our Favorite Flavors of Ice Cream

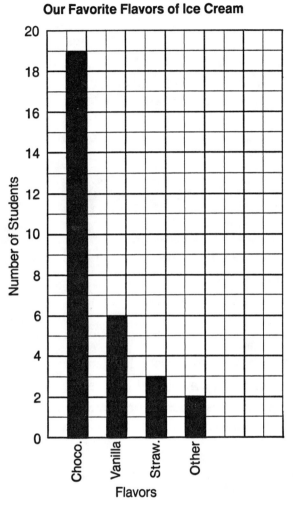

Figure 5-4. Graph: Our favorite flavors of ice cream

4. Analyze the data on the completed graphs, discuss:

   a. What is the favorite flavor of ice cream in our class?

   b. What is the second favorite flavor of ice cream?

   c. If we were to order ice cream for a class party, what flavors should we order?

   d. What flavors of cake do you infer are favored by the class?

5. Display some of the graphs.

"If You Like" Assignments

Follow Information Sheet 5-A, "Guidelines for Conducting an Independent Survey," such as:

- What is the most popular pet in our class?
- How many hours a day do you watch TV?

Lesson 3. Animals Who Visit Our Bird Feeder

OBJECTIVES

Students will:

1. Set up and stock a bird feeder.
2. Record, on a table of data, observations of animals who visit the feeder.
3. Transfer the data to a bar graph.
4. Interpret the graph and the data.

MATERIALS

Student Worksheet 5-1, "Animals Who Visit Our Bird Feeder"

Feeding station that can consist of bird seed:

- sprinkled on a windowsill, ledge, or plank
- sprinkled on the ground
- placed in a feeder

A bird watching station can be made of a desk and chairs placed by a window with:

- field guides for identifying birds and small mammals
- binoculars (optional)
- pencils, colored pencils, crayons, and drawing paper

Also:

- paper for individual tables of data
- graph paper for individual graphs

INSTRUCTIONAL PROCEDURES

Step 1. Identifying the investigation.

Explain that the class is going to set up a bird feeder, keep it stocked with bird seed, keep records of the birds and other animals that visit it at a certain hour each school day, and graph the results.

Guidelines for Conducting an Independent Survey

There are many different ways to conduct a survey. Listed below are the steps that you can follow to survey and report on the results.

1. Think about what you want to investigate.

2. Identify the group of people that you wish to question. That is, are you interested in the answers of children, adults, or some other specific group?

3. Form a question that when answered, will give you the information you are seeking.

4. Make a table of data. Use your question as the title.

5. Ask your question and record the results on your table of data.

6. Transfer your data to a graph. For the title, use the question or a descriptive phrase explaining what you discovered.

7. Analyze the data on the table and the graph to answer your question.

8. Write brief statements:

   a. What you investigated.

   b. Your procedures.

   c. Your findings.

9. Put your statements, table of data, and graph together in a report.

10. Bring your report to class to give a presentation and/or to set up a display.

Name _____ **Date** _____

Animals Who Visit Our Bird Feeder

The number and kind of animals who visit our feeder at _____ A.M. during the week

of _____.

Day 1	Day 2	Day 3	Day 4	Day 5

- Set up the feeder.
- Make the table of data.
- Record the observations for Day 1.

Step 2. Gather and record the information on a table.

Identify and record the types and number of animals on the table of data, such as, four Blue Jays, two squirrels. Discuss the day's observations.

Step 3. At the end of the week, direct students to transfer the findings to a bar graph.

While students work, check their graph to determine if they:

- Use a base line. Write words under the base line. Record data above the base line.
- Write numbers on lines, not spaces.
- Label each bar.
- Write a zero at the intersection of the two axes.
- Label the horizontal axis with names of animals, such as "Blue Jays, Cardinals, Squirrels, Other."
- Label the vertical axis, "Number of Animals."
- Write a title.

Step 4. Interpret the graph.

Direct students to use their graphs to answer the following questions:

- Who visited our feeder?
- Did only birds visit our feeder? Why do you think that happened?
- How many animals visited on the first day?
- On which day did we have the most visitors?
- Why do you think we had more visitors on that day?
- What do you predict regarding the number of animals who will visit the feeder tomorrow?
- What do you predict about the animals who will visit the feeder next week?
- How confident are you in your predictions?

Step 5. Present your findings.

Present the findings by displaying the table of data and graphs on a bulletin board. Explain to your class that scientists present their tables and graphs along with a formal presentation at conventions and in articles. These scientific presentations usually include statements of purpose, findings, and recommendations along with the tables of data and graphs.

"If You Like" Assignment

Set up a feeder of your own at home.

Teaching Tips

1. Distribute copies of Information Sheet 5-B, "Guidelines for Conducting an Investigation at Home" and Information Sheet 5-C, "A Letter to Parents."
2. Review the information on both sheets with the class but make the assignment optional.
3. Use the "Letter to Parents" for investigations at home.

Lesson 4. Introducing Line Graphs (Grades 5–8)

OBJECTIVES

Students will:

1. Transfer information from a table of data to a line graph.
2. Explain that the horizontal or base line is the manipulating axis, and that which the investigator manipulates is written along the manipulating axis.
3. Explain that the vertical axis is the responding axis, that which responds to the investigation, and is recorded along the responding axis.
4. Make extrapolations based on predictions which "exit" the graph.

MATERIALS

Overhead projector
A grid transparency
Graph paper
Ruler

ON THE BOARD/Tables of Data

Maze Time for Mazie

Trial	Time in seconds
1	35
2	15
3	10
4	5
5	5

Maze Time for Stacy

Trial	Time in seconds
1	60
2	30
3	20
4	10
5	5

Name _____ Date _____

Guidelines for Conducting an Investigation at Home

Before you investigate:

1. Give your parents or guardian the letter stating that you are interested in conducting an investigation.

2. Explain to them what you want to do and how you want to do it.

3. If they approve, discuss safety guidelines, cost, and equipment that you may use.

4. Never use your parents' equipment or tools without their permission and directions for proper and safe use.

5. Ask one parent or guardian to sign the approval form and return it to me.

6. Assemble all of the equipment that you need.

Your Investigation

In conducting your investigation, make sure that you understand the following steps. You may add more if you like, and may decide whether or not to complete an "optional" step. You may conduct the investigation for any number of weeks that you like, but be realistic.

STEPS

1. In writing, briefly state the purpose of your investigation.

2. Make a table of data on which you will record your observations. Include a descriptive title which can later be used for the title of your graph.

3. At the end of each week, graph your data.

4. Check the accuracy of each bar.

5. As a part of your observations, take photographs and/or make sketches. (Optional.)

6. Use binoculars. (Optional.)

7. Use reference books and/or field guides.

8. Write a brief summary stating what you learned.

9. Put your presentation together. It should include:

   a. Statements of purpose and procedures.

   b. Tables of data.

   c. Graphs.

   d. A summary of what you learned.

   e. Sketches and photographs. (Optional)

10. Bring your presentation to class. Give report and/or set up a display.

A Letter to Parents

_____ School

_____ Date

Dear Parents:

Your child has expressed an interest in conducting a voluntary investigation or project at home. To do this he/she must:

- get your approval, and
- work out with you all of the details including cost, equipment, and safety regulations.

An information sheet is attached to help you in your decision making.

If you have any questions, please call me at school and leave a message, including your phone number where I can reach you during the school day. I will return your call as soon as possible.

Sincerely,

Teacher

Administrator

- -

I give my approval for my child _____

to _____.

I have reviewed the directions and safety procedures required in my home.

_____

Parent's Signature

INSTRUCTIONAL PROCEDURES

Step 1. Read the following graphing problem to the class.

Mr. Foley's sixth grade science class has a pair of white mice named Mazie and Stacy. The students wanted to determine if the mice could learn their way through a maze. They built two identical mazes out of cardboard boxes, one for Mazie and one for Stacy. Then each day for the next five days they put two slices of fresh carrot at the end of each maze, put each mouse in its maze, and timed the number of seconds that it took each mouse to reach its food. They recorded the time for each trial on two tables of data.

Step 2. Refer to the tables of data on the board.

Step 3. Transfer the data to graphs:

A. Demonstrate how to transfer the data from "Maze Time for Mazie" onto a line graph.

   (1) Explain that whatever the investigator controls or manipulates is written along the horizontal or "bottom" axis, which is called the manipulative axis.

   (2) Explain that whatever element responds is written along the vertical or "side" axis, which is called the responding axis.

   (3) Demonstrate that on a line graph a point showing the coordinates takes the place of the top of a bar on a bar graph.

   (4) Use a ruler to connect each pair of consecutive points.

   (5) Explain that line graphs are used to show growth and decline.

B. Direct students to graph the data from "Maze Time for Stacy."

Step 4. Interpret the graphs.

A. Compare the graphs to answer such questions as:

   (1) Did the mice learn to go through the maze? (Yes)

   (2) Which mouse learned first? (Mazie)

   (3) What do you *predict* the time in seconds will be for a sixth trial for Mazie? (Five seconds or less.)

   (4) For Stacy? (Five seconds or less.)

B. Explain that a prediction made beyond the actual points plotted on the graph is called an "extrapolation." Extrapolations are an important part of science which enable scientists to make predictions.

C. List some inferences:

   (1) Female mice are smarter than male mice.

   (2) Female mice like carrots better than male mice.

   (3) Mazie can smell better than Stacy.

Teaching Tip

If you have mice, gerbils, or guinea pigs, consider investigating their maze time.

Lesson 5. Graphing the Melting Time of Ice Cubes.[1]

OBJECTIVES

Students will:

1. Construct a line graph given a table of data.
2. Interpret the findings and the graph.

MATERIALS

Student Worksheet 5-2, "What Is the Melting Time of Ice Cubes?"
Graph paper
Rulers

INSTRUCTIONAL PROCEDURES

1. Distribute the worksheet.
2. Review the directions.
3. Distribute graph paper and rulers.
4. When students have completed the worksheet, review the answers.

   a. After students make an interpolation by correctly answering questions two and three, introduce the term "interpolation" as interpreting data that falls inside the plotted coordinates.

   b. After students make an extrapolation by correctly answering question four, review the term "extrapolation" as interpreting data that "exits," or falls outside of, the data.

---

[1] Science—A Prcocess Approach/D. Predicting 4/The Suffocating Candle, American Association for the Advancement of Science/Xerox Corporation, 1968.

Name _____ Date _____

What Is the Melting Time of Ice Cubes?

Two students were interested in the time it takes ice cubes to melt when placed in water and left at room temperature. First, they filled two large peanut butter jars halfway with water, which they let stand until it reached room temperature. Next, they added four ice cubes to one jar and eight ice cubes to the other jar. They measured and recorded the amount of time that it took for each set of ice cubes to melt.

Table of Data

Melting Time for Ice Cubes in Water at Room Temperature

Number of Ice Cubes	Melting Time in Minutes
4	20
8	60

Directions:

Transfer the data to a line graph, answer the following questions, and write two questions of your own based on the investigation.

1. How do you determine on which axis to place each set of information?

2. How long will it take six ice cubes to melt?

3. How long will it take seven ice cubes to melt?

4. How long will it take nine ice cubes to melt?

Answers to Student Worksheet 5-2,
"What Is the Melting Time of Ice Cubes?"

Melting Time for Ice Cubes in Water at Room Temperature

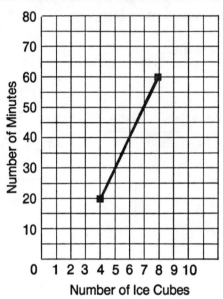

Answers to the Questions

1. The investigator controls or manipulates the number of ice cubes in this investigation. Thus, "the number of ice cubes" is written along the horizontal, or manipulating, axis. "The number of minutes" that it takes the ice to melt is the response and is written along the vertical, or responding, axis.

2. 40 minutes.

Explain that since this inference comes from "inside" the data plotted on the graph, it is called an interpolation. An easy way to remember this is to recall that "interpolate" and "inside" both begin with "in."

3. 50 minutes.

4. 70 minutes.

Explain that since it is based on data that "exits" the plotted coordinates, it is called an extrapolation. It helps to remember that "extrapolation" and "exit" both begin with "ex."

Lesson 6. Analyzing a Completed Graph

OBJECTIVES

Students will:

1. Use a completed graph to make an economic decision.

2. Identify another way to measure the growth of plants.

3. Make an extrapolation based on the graph.

4. Design a similar investigation to determine if they are getting their money's worth.

MATERIALS

Student Worksheet 5-3, "How Does My Garden Grow?"
Student Information Sheet 5-D, "The MacGregors' Garden Graphs"
Graph paper
Rulers

INSTRUCTIONAL PROCEDURE

1. Place students in working teams of two or four.

2. Distribute both worksheets.

3. Review all of the directions.

4. Distribute the graph paper and rulers.

5. When students finish, review the answers to the worksheet.

Answers to Worksheet 5-3, "How Does My Garden Grow?"

1. Yes. The plant that received the plant food grew faster.

2. They could have measured:

- the weight of the plant
- the width of the foliage
- the number of flowers

3. Plant with fertilizer _____10 cm_____.
   Plant without fertilizer ___8 cm_____.

4. Yes. If the plant food helped the geranium grow faster, it might help other plants grow faster as well.

Teaching Tips

Turn this lesson into a lab:

1. Use plant cuttings. African violets, pussy willows, and forsythia make excellent cuttings for classroom use.

2. Use seeds. Lima bean, radish, and sunflower seeds grow quickly, thus making excellent plant investigations.

Name _____   Date _____

How Does My Garden Grow?

Mr. and Mrs. MacGregor wanted to determine if the plant food they were using was worth the cost. They took two similar size cuttings from a large geranium plant and planted each cutting in a clay pot full of potting soil. Both plants were kept in the same conditions with one exception: one received plant food according to the directions on the package and the other received no plant food. The MacGregors, who take their gardening and finances seriously, recorded the measured height in cm of each plant for four weeks on a table and transferred that information to two graphs.

Directions: Use the graphs to help you answer the following questions.

1. Do you think that the MacGregors decided that the plant food was worth the expense?

    _____ Why do you think as you do?

    _____

    _____

    _____

2. What else could they have measured besides the height of the two plants? _____

    _____

    _____

3. What do you predict will be the height of each plant during week five?

    Plant with fertilizer? _____

    Plant without fertilizer? _____

4. If you were the MacGregors, would you try the plant food on other plants in addition to

    geraniums? _____

    Why? _____

    _____

    _____

    _____

    _____

Name _____   Date _____

The MacGregors' Garden Graphs

Plant with Food   **Plant without Food**

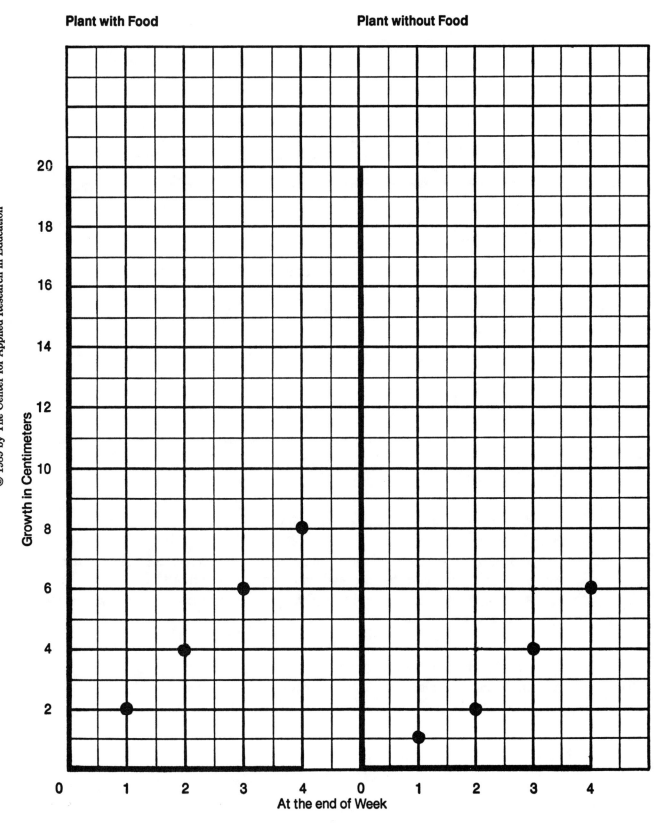

At the end of Week

Lesson 7. Do Preservatives Prevent Mold on White Bread?

(Culminating Graphing Experience For Grades 5–8)

OBJECTIVES

Students will:

1. Set up an investigation of the effect of preservatives on bread mold.
2. Make daily observations and record the results on tables of data.
3. Transfer the results to graphs.
4. Analyze the graphs.

MATERIALS

On each tray:

      1 piece of white bread with preservatives

      1 piece of white bread without preservatives

      2 clean petri dishes or 2 plastic bags (self-locking or with ties)

      1 transparency of a grid made from graph paper

      ¼ cup of water

Paper for tables of data

Graph paper

Student Worksheet 5-4, "Do Preservatives Prevent Mold on White Bread?"

INSTRUCTIONAL PROCEDURES

Step 1. Identify the investigation.

To determine if preservatives can prevent the growth of bread mold on white bread:

A. Identify one piece of white bread that has preservatives in it and one that does not by reading the information on the bread wrappers.

B. Treat both pieces of white bread equally. Scientifically we say that we are controlling the variables:

(1) Dampen each piece of bread.

(2) Wipe the dampened side on a counter or the floor.

(3) Place each piece in a clean petri dish.

(4) Cover with the lid.

Name _____   **Date** _____

Do Preservatives Prevent Mold on White Bread?

Day	Bread without preservatives — Number of squares with mold	Bread with preservatives — Number of squares with mold	Conclusions
1			
2			
3			
4			
5			
6			
7			
8			
9			
10			
11			
12			

For the weekend, record, "No observation."

Step 2. Gather and record the data on an organized table.

A. Each day, at the same time, for the next two weeks, observe the bread and record observations:

(1) Place the grid transparency on the closed petri dish containing the bread *without* the preservatives.

(2) Count the number of squares that have mold growing under them.

(3) Record that on the table of data.

(4) Repeat the same procedure for the bread *with* the preservatives.

Step 3. Graph the data.

Each student transfers the data from the table to two graphs:

- Bread with preservatives
- Bread without preservatives

Step 4. Interpret the graph.

Use the graphs to discuss the following questions:

1. Do preservatives prevent mold from growing on white bread?
2. Do preservatives retard, or slow down, the growth of mold?
3. Did you observe more than one kind of mold?
4. What do you predict would happen if we conducted this experiment on other kinds of bread?
5. What do you predict would happen if we conducted this experiment on cakes?
6. What is another name for the horizontal axis?
7. What is another name for the vertical axis?
8. Explain how to determine which set of data goes on which axis?

Step 5. Present the findings

Display the tables of data and the graphs.

Teaching Tip

1. Conduct this investigation in class with other baked goods.
2. If you are concerned that too many of your students have allergy problems to do this investigation, teach it as a demonstration.

"If You Like" Assignments

1. What is the effect of preservatives on rye bread?
2. What is the effect of preservatives on potato chips?

3. What is the effect of preservatives on vanilla cookies?

4. What is the effect of refrigeration on mold?

5. What is the effect of darkness on mold?

For Your Consideration

This type of an investigation lends itself to formal presentations. It is ideal for science fairs, classroom presentations and school displays.

Thus, after completing this investigation, you have an excellent opportunity to show students new ways to present their findings. Consider introducing the three-fold panel.

Lesson 8. How to Make a Three-Fold Panel

OBJECTIVES

Students will:

1. Explain how to make a three-fold panel.

2. Explain how to use a panel for presentations in class and in science fairs.

MATERIALS

Information Sheet 5-E, "'Making a Three-Fold Display Board"

Information Sheet 5-C, "Letter to Parents."

Three rectangular pieces of cardboard, plywood, or pegboard

Four hinges

Eight nuts and bolts that are compatible with the hinges

Pliers

Masking tape

Seven rectangular strips of oaktag or manila paper

Felt tip markers

Before Class/Make a Three-Fold Display Board

Either with a helpful hand at home or a student volunteer at school, assemble the three-fold display board:

1. Have someone hold two of the pieces of board together while you bolt the hinges in place.

2. Use the oaktag strips and markers to make the seven title strips, as is shown in the diagram of a sample three-fold display board.

Name _____   Date _____

Building a Three-Fold Display Board

This set of guidelines will help you construct a display board. Read all of the directions first. Think about the component parts of your investigation and how you can best display them. Be creative. Use your imagination.

A Sample Board

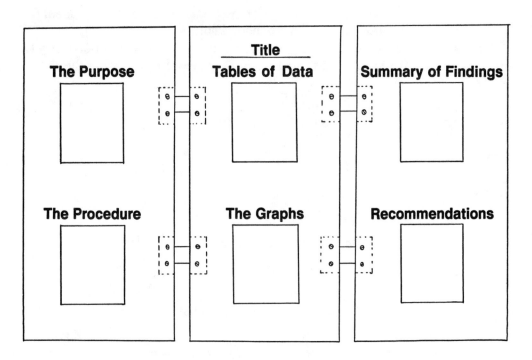

Before you begin to build:

1. Give your parents or guardian the letter stating that you are interested in this voluntary project.
2. Explain to them what you want to do and how you plan to do it.
3. If they approve, discuss safety, equipment, and finances.
4. Have them sign the approval sheet so you can return it to me.
5. Assemble all of your supplies.

Materials:

1. For the three panels, use cardboard, plywood, or pegboard.
2. To hold them together use a strong tape, shoe laces, yarn, or hinges, nuts and bolts.
3. To make the board more attractive, make easy-to-read title cards and attractive art work.

Measurements:

Minimum measurements for one panel, 44″ by 15″.

Each panel needs to hold at least two sheets of standard paper each measuring 8½″ by 11″ and two or three title cards each about three inches tall. Thus, the minimum measurement for one panel should be 44″ by 15″. The middle panel may be wider than the end panels, but not taller.

If you are entering a science fair, ask someone in charge of the science fair for the size requirements or restrictions for displays.

INSTRUCTIONAL PROCEDURE

Demonstration

1. Show the display board.
2. Explain that they are excellent for making formal presentations of scientific investigations.
3. Demonstrate:
   a. Folding the board into thirds for storage and easy transportation to and from school.
   b. Standing the board on its own when opened.
   c. Making the board by unfastening a hinge. Show the hinge, bolt, and screw.
   d. Taping the seven title strips on the board. Ask for other ways to put titles in place . . . velcro, pockets, ribbons.
   e. Putting a piece of paper containing information in place . . . use masking tape, or make clear plastic pockets into which one slips the paper.

DISCUSS

1. An excellent and inexpensive display board can be made out of cardboard and masking tape. It can be as effective for presentation purposes, but it won't be as durable.
2. The use of spray paint and/or contact paper will make it more attractive.
3. The written parts must be clearly expressed and neatly presented . . . in ink, typed, or computer printout.
4. Tables and graphs must be correctly and neatly labeled.
5. The math and the science must be accurate.
6. The materials used in the investigation should be displayed in front of the opened board.
7. If this is for competition in a science fair, each component needs to be artistically presented.
8. Students may borrow the class board for a two-week period for voluntary class presentations by signing up.
9. Information sheets are available at any time for those who are considering making a display board.

The Snowball Effect

After the first volunteer brings a project to class and explains the investigation, class interest increases, and more projects are brought to class. There is a "snowball" effect.

Teaching Tips to Consider When Teaching Graphing

- Demonstrate how and where to place some of the data on a graph.
- Follow your demonstration by having students construct the graph that you started and complete it using the available data.
- Make it relevant by providing concrete examples.
- Make it interesting by using a variety of hands-on materials.
- Provide lots of practice with bar graphs before introducing line graphs.
- Use graph paper that is easy on children's eyes. Graph paper with squares measuring 1 square cm or ½ square cm is recommended.
- Teach students to identify investigations that lend themselves to graphing, such as surveys and experiments, in which they manipulate a variable and record the response.
- Continue to use the graphing skills, even after the graphing lessons are over, by making graphing an integral part of other investigations.
- Transfer the skills for interpreting graphs to other areas of study, such as interpreting the meaning of graphs in social studies books and reading workbooks. Use pamphlets from health and environmental agencies. This will demonstrate that the non-school world really uses graphs.

A Note of Special Interest

My special education students—those identified as Learning Disabled and Socially and Emotionally Maladjusted—did as well with graphing as my other students, including those identified as Gifted and Talented.

For More About Graphs

For further graphing experiences for students in grades 3–8, I recommend the program *Science—A Process Approach,* developed by the American Association for the Advancement of Science. In the 1968 edition, Levels D and E contain many graphing lessons.

The Human Body:
A Cut-and-Paste Model

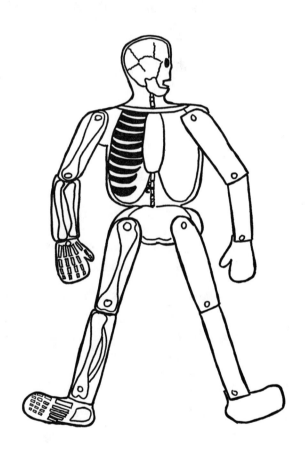

Personal Perspective

A review of newspapers, magazines, and TV programming reveals interest and problems with general health, diet, alcohol, and drugs. More people than ever are interested in improving or maintaining good health through a personal fitness program. Sales of health club memberships, walking shoes, and diet dinners are up. As educators, we can capitalize on this interest to teach our students about the body: how it works, what it needs to be healthy, and how to use this information to develop habits which promote individual health and safety.

Since the internal workings of the human body can be deceptively abstract, it is important to develop lessons that provide students with concrete experiences. Making a model of the human body provides students with opportunities to identify shape, placement, and function of internal organs and systems in the body. When students understand the workings and interrelations of five of the body's systems presented in this section, they then can understand the answers to many of their own questions such as:

1. Do I really need to drink milk?
2. What's wrong with trading my sandwich for cookies or potato chips?
3. What's wrong with smoking cigarettes or pot?
4. What do drugs do? Why are they so popular?

A Highly Prized Project Provides Motivation

"My brother was in your class two years ago and he still has his skeleton hanging in his room." That's a phrase I frequently hear when I introduce our investigation of the human body by showing a completed cut-and-paste, oaktag model of the human body. The "skeletons" as the students call them become highly prized and survive for years.

These "skeletons" consist of an oaktag silhouette held together at twelve joints by brass paper fasteners. One side has a drawing of the skeletal system. The other side has cutouts of the brain, heart, digestive system, and respiratory system. Movable joints and a pair of lungs which can be lifted to reveal other parts are two features which attract students' attention.

As students make "paper doll" models of a system, they:

1. See what the organs look like and where they are located in the body.
2. Discuss the function and needs of each organ and system.
3. Learn about the effects of food, exercise, rest, pollution, alcohol, and drugs upon an organ and system.

166

4. Discuss making decisions depending on how the outcome can affect the body.

5. Identify personal habits that are conducive to promoting the healthiest body possible.

Before Teaching This Unit

1. Locate and organize all of the art supplies.

   - Oaktag for body silhouettes
   - Unlined paper for tracing organs and systems
   - Scissors
   - Felt tip markers
   - Crayons or colored pencils
   - Glue, small cups, and popsicle sticks

2. Identify storage places for art supplies and partially completed student models of the human body.

3. Make one pattern of the human body silhouette for every group of four to six students. Use the patterns provided on pages 168, 169, and 171.

   - Laminate the patterns and use them for years to come.
   - Organize the pattern pieces by placing each complete pattern set, 13 pieces, in a large envelope and label it.

4. Make and laminate the patterns for the five body systems. Use the patterns provided in this section.

   - Organize the patterns. Place all of the patterns for one system in an envelope and label it.
   - If you prefer, make one copy of page 170 for each student. As the class investigates each system, direct students to color, cut, and paste the system on the drawing.

5. Make one complete model of the human body with all five systems for use in your introductory lessons and for motivation.

6. Identify chapters in science or health books about the human body and nutrition for use by students and substitute teachers.

7. Start a bone collection. Contact local butchers; ask for long bones, ribs, and a backbone. Ask to have a long bone cut to show that it is hollow inside. Save and clean bones from meals. Store the collection and add to it over the years. A skull is a good attention getter.

8. Locate and display posters of the body systems and foods.

9. Ask a local radiologist for x-rays of hands, feet, legs, skull, arms, and chest cavity. Hang them in classroom windows so the light can pass through them.

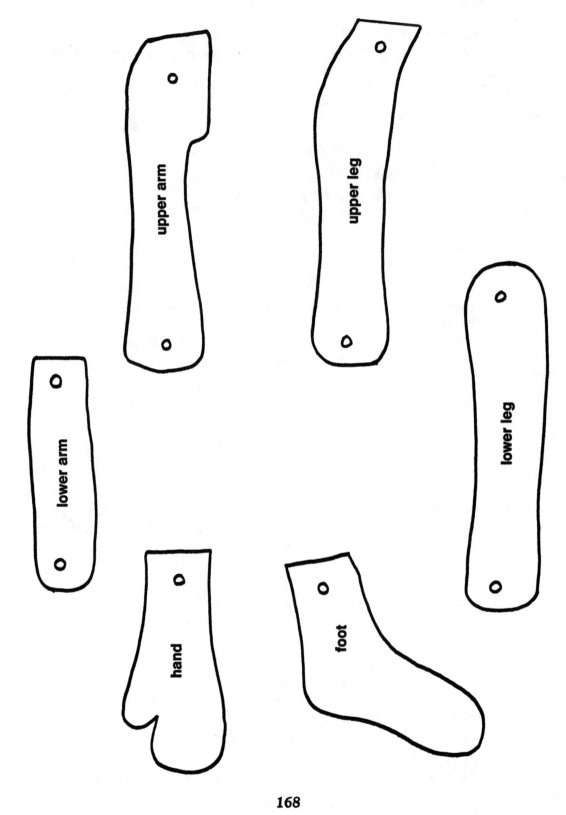

168

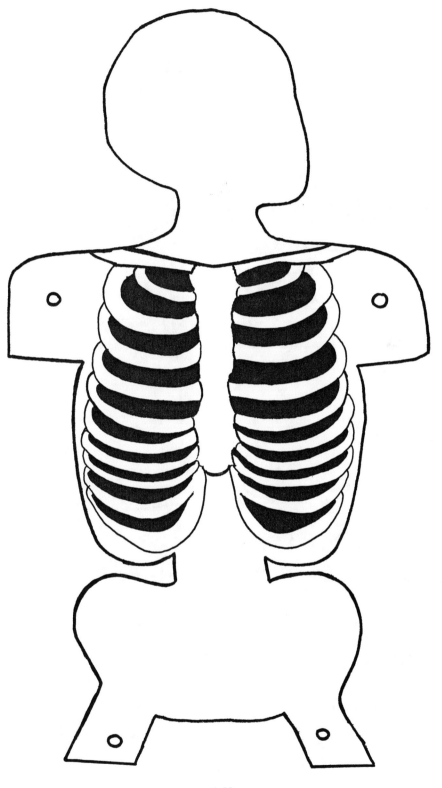

169

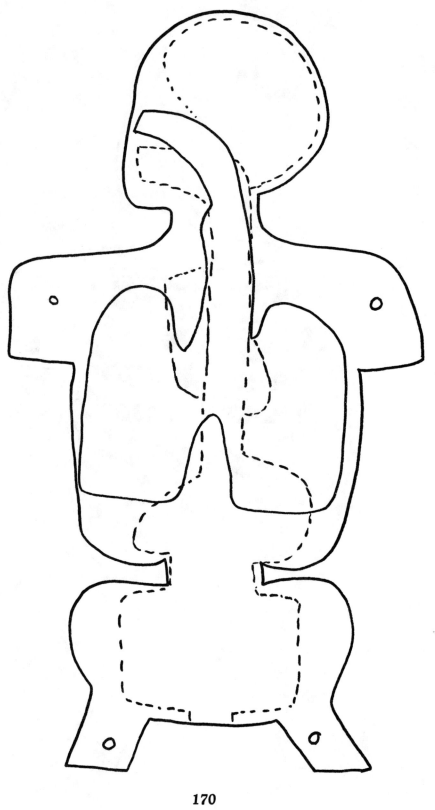

170

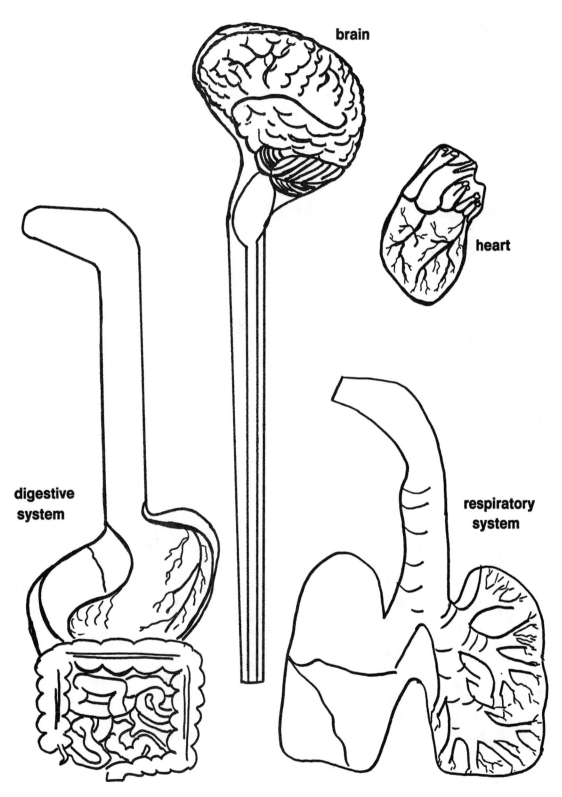

brain

heart

digestive
system

respiratory
system

171

Preparing to Teach This Unit for the First Time

In preparing to teach this unit for the first time, set realistic goals for yourself and your students. Concentrate on understanding the five systems covered in this chapter. Select lessons that are appropriate for your class and duplicate the worksheets. Plan to add to your knowledge, collections, and activities a little each year. Younger students need at least two class periods for cut-and-paste sessions. It takes me six to eight weeks to teach this unit.

Recommended Steps for Teaching This Unit

Use a completed cut-and-paste model of the human body to introduce this unit and each of the five systems. Teach about one system at a time. To do so, the following steps are recommended:

1. Identify the name and major function of the system.
2. Identify the major organ and its function.
3. Discuss possible consequences of good and poor health habits upon the system.
4. Provide at least one activity that will enhance students' understanding of the system.
5. Have students practice sketching a system before adding a drawing of it to the model of the body.
6. Have students add a drawing of a system to the model of the human body as the culminating activity for each system.

Lesson 1. Introducing the Human Body Unit and the Skeletal System

OBJECTIVES

Students will:

1. Explain that each student will make a model of the human body.
2. Explain that the major functions of the skeletal system are to support and protect the body.
3. Explain that bones move at joints with the help of muscles.
4. State that long bones make up the arms and legs.
5. State that short bones make up the fingers and toes.

MATERIALS

A completed cut-and-paste model of the human body

A poster or model of the human skeleton

A long bone, rib, backbone, shoulder blade, and skull (optional)

INSTRUCTIONAL PROCEDURES

Introduce the Unit

Show students the cut-and-paste model of the human body in such a way that they will want one. To do this, show the side with the skeleton drawn on it. Hold it by the neck and sway it to make the joints move. Turn the model around to show the five systems on the other side. Lift the lungs to display the systems underneath.

As you show the model, explain that everyone will be given the supplies, patterns, and directions that will enable each person to make his or her own model of the human body. Also explain that it takes about six to eight weeks to learn about the body and complete the model. The model will be stored in the classroom until it is completely finished and evaluated.

Also, explain that the model will be evaluated on whether or not the directions are followed.

Introduce the Skeletal System

Show the skeletal side of the model, and ask, "What would we be like without a skeleton?" (Jellyfish, worms) Listen to their ideas. Discuss the major functions of the skeleton:

1. It provides the framework for our body.
2. It provides protection for internal parts.
3. Along with our muscles, it helps us move.

Student Activity

Direct students to:

1. Bend the elbow. (Can be done.)
2. Keep the elbow bent and bend the wrist. (Can be done.)
3. Bend the arm between the wrist and the elbow. (Cannot be done.)

Explanation

We cannot bend the arm between the wrist and the elbow without breaking the arm. There are two long bones which go from the wrist to the elbow. Bones cannot bend. Bones move at the joints, the meeting of two bones, with the help of the muscles. The elbow and the wrist are examples of two types of joints. Thus, we can bend at the elbow and the wrist, but not in between.

Student Activity

Direct students to:

1. Stand up.
2. Place one foot on their chair.
3. Place one hand on their own knee and another on their own hip.
4. Bend the leg somewhere between the two hands. (Cannot be done.)

Explanation

The upper leg has one long bone. The bone does not bend. The leg can move at the knee and hip which are joints. The knee and the elbow are hinge joints. They move like the hinge on a door. The shoulder and hip are ball and socket joints. They move like a ball and socket pen holder or a shower head.

Teacher Demonstration

Hold a long bone on your upper arm and upper leg to show where the long bone is and how it is shaped. If you do not have a large long bone, identify long bones on a poster.

Student Activity

Tell students to make a "C" with the index finger and thumb; and count the number of short bones in each finger and thumb. See Figure 6-1.

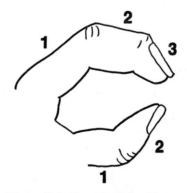

Figure 6-1. Counting short bones

Explanation

There are two short bones in the thumb and three short bones in each finger. We have an opposable thumb. That is, the thumb can move opposite to any of the fingers. This allows us to use tools and make fine movements with our hands.

"If You Like" Assignments

1. Take off your shoes and socks and count the short bones in each toe.
2. Try to pick up objects or to write without using your thumb.

Lesson 2. Students Observe and Sketch Bones

OBJECTIVES

Students will:

1. Observe real bones or pictures in texts.
2. Sketch long and short bones, a back bone, a skull, and a rib cage.
3. State that long bones are hollow.
4. State that red blood cells are formed in rounded ends of long bones.

MATERIALS

Pictures of the skeleton
One long bone cut open
Bones or pictures of bones for sketching
At least one long bone (chicken leg) for each group

INSTRUCTIONAL PROCEDURES

Hold up one of the leg bones from a chicken, ask, "What type of a bone do you think this is?" (Long bone)

Explanation

Long bones have a characteristic shape: rounded ends with cylindrical middles. This is the long bone of a chicken.
Show the class the bone that has been cut open. Ask for observations. (It is not solid bone.)

Explanation

Most bones are hollow. This makes them lighter and thus makes it easier for us to walk, run, and jump. Point out that the ends of the long bones are extremely important to us because that is where our red blood cells are made in the red marrow.

Student Activity

Give students as many bones as possible, but at least one long bone of a chicken. Direct them to:

1. Sketch the bones.
2. Refer to posters and books to practice sketching parts of the skeleton.
3. Pass around the opened bone to observe the inside of a bone.

"If You Like" Assignments

1. Make a drawing of the human skeleton and label some of the 206 bones. Use dictionaries, encyclopedias, and science books.
2. Clean and sun bleach the bones from a turkey or chicken dinner and label some of the bones; rib and long bones are easy to identify.

Lesson 3. The Importance of Minerals to Bone

OBJECTIVES

Students will:

1. Describe the hardness of a bone that is rich in calcium and other minerals based on observations.
2. Describe the lack of hardness in a bone that has lost some of its minerals.
3. Identify at least two sources of calcium that they could consume on a daily basis to help develop strong bones and teeth.

MATERIALS

One long bone from a chicken for a control
On each tray:
    One long bone of a chicken (Try other bones.)
    A small jar and cover (Use plastic wrap if you don't have a lid.)
    Enough vinegar to cover the bone

INSTRUCTIONAL PROCEDURES

Lab or Demonstration

Direct students to:

1. Write at least four observations of the chicken bone.
2. Place the bone in the jar and cover it with vinegar.
3. Cover it to prevent the vinegar from evaporating.
4. Remove the bone from the vinegar after four days and rinse it with water. Write four observations.
5. Compare both sets of observations.
6. Compare the experimental bone with the control bone.

Explanation

Vinegar chemically removes some of the minerals from the bone. When a bone is poor in minerals, it loses some of its hardness, becomes weak, and is more susceptible to breaks. Calcium and phosphorus are two minerals that form hard proteins in bones and teeth.

Our body needs calcium to function. If it doesn't get enough from the daily diet, it will take from the bones. Thus, it is important to consume calcium and other minerals on a daily basis.

Milk and milk products are rich in calcium and phosphorus. There are calcium tablets for those who are allergic to or cannot digest milk products.

While a person is growing he or she is forming bones and teeth. If he doesn't get enough calcium, he will still form teeth and new bone but the bones and teeth will not be as strong or as healthy as they could be. The United States Department of Agriculture recommends the following number of daily, eight ounce servings of dairy products:

Children . . . three to four

Teen-agers . . . four or more

Adults . . . two or more; low fat.

Teaching Tips

1. If you are concerned that the vinegar may spill on the patterns or student models, do this lesson as a demonstration.

2. Discuss the importance of students being responsible for making healthy food choices, especially when parents are not present. For example, if they have not had enough food containing calcium in a day, they should not choose soda in place of milk at a meal.

"If You Like" Assignments

1. Make a poster or mobile of milk products.
2. Learn to make a milk snack, such as a milk shake, eggnog, or pudding. Cooking and operating appliances requires adult approval and assistance.
3. Keep a daily diary of what you eat for a week. At the end of each day, record and count the number of servings of milk products eaten that day.

Lesson 4. What Does It Mean to Stand Up Straight?

Note: This is an optional lesson for students in grades 6–8.

OBJECTIVES

The students will:

1. State that good posture is important.
2. Explain that good posture means an "S" curve to the backbone.
3. Identify two habits that they can develop which will promote good posture.

MATERIALS

A model or picture of the backbone

Vertebrae or a spool of thread to be used as a model of vertebrae

Five to ten books

Student Worksheet 6-1, "Looking at the Backbone"

INSTRUCTIONAL PROCEDURES

Minilecture

Tell students to think about what it means to stand and sit up straight. Show a model or picture of the backbone. Ask if it is perfectly straight. (No)
Ask for descriptions.

1. A "straight" backbone has an "S" shape.
2. The backbone is not one bone but a column of bones stacked on top of each other.
3. All of the vertebrae are not the same size.
4. Each vertebra has a hole in the center. (Show the model or the spool of thread.) These openings provide a protective tunnel for the spinal cord.

Name _____   Date _____

Looking at the Backbone

Directions: Label the parts for the two diagrams. To do so, match the number in the drawing to the number in the list.

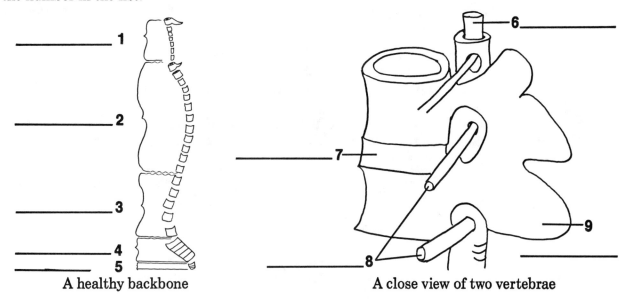

A healthy backbone   A close view of two vertebrae

1. Seven Cervical vertebrae
2. Twelve Thoracic vertebrae
3. Five Lumbar vertebrae
4. Sacrum
5. Coccyx

6. Spinal cord
7. Disc between vertebrae
8. Spinal nerve
9. Spines of vertebrae

Directions: Use the information in the two diagrams to answer the questions listed below.

1. Should the backbone be perfectly straight?   (Yes or No)

2. The shape of the backbone should resemble the letter _____.

3. The backbone is made up of _____ bones.
                                  (number)

4. There are _____ cervical vertebrae in the backbone.
          (number)

5. There are _____ thoracic vertebrae.
          (number)

6. There are _____ lumbar vertebrae.
          (number)

7. There is _____ sacrum.
          (number)

8. There is _____ coccyx.
          (number)

9. The opening in the center of each vertebra provides a continuous protective tunnel for the

_____.

5. The spinal cord is the central pathway for all messages to and from all parts of the body. Damage to it is dangerous because it can result in paralysis or death.

6. We need strong vertebrae stacked on top of each other in an "S" formation to provide protection for the spinal cord.

Next, pick up one book at a time. Hold the books in the normal way on one side of your body. Deliberately allow that shoulder to drop, demonstrating that carrying books in one arm can lead to poor posture. Ask students for observations. (Problems with posture)

Ask what they can do about it. (Getting rid of books is *not* an acceptable answer. Alternating the arm in which books are carried or carrying books in back packs are acceptable answers.)

Student Activity

Students work in teams to answer Student Worksheet 6-1, "Looking at the Backbone."

Answers to Worksheet 6-1, "Looking at the Backbone "

1. No	4. 7	7. 1
2. "S"	5. 12	8. 1
3. 26	6. 5	9. spinal cord

FOLLOW-UP DISCUSSION

Identify actions that students can take to protect their own skeletal system, such as:

1. Eat four servings of dairy products every day or take calcium and phosphorus supplements.

2. Alternate carrying books in different arms.

3. Use a back pack.

4. Work on good posture for walking, sitting, and standing.

5. Think "safety" when playing.

6. Wear protective head gear and clothing when participating in sports.

7. Never substitute a soda for a glass a milk.

Teaching Tips

1. Invite a physical therapist or chiropractor to bring a model of the backbone to class and discuss the importance of food, posture, and diet.

2. Do not require students to memorize the names or numbers of vertebra. Require them to explain that the backbone is made of small bones stacked on top of each other to make the form of an "S."

Lesson 5. How to Put the Pattern Together and Where to Draw the Bones

OBJECTIVES

Students will:

1. Explain that making the cut-and-paste model is like making a puzzle and then putting the pieces together.
2. Explain that the paper fasteners hold two pieces of paper together to form a model of a joint.
3. State that the skeletal system is drawn on one side of the silhouette.

MATERIALS

Worksheet 6-2, "Preparing to Make a Model of the Human Skeleton"

INSTRUCTIONAL PROCEDURES

Distribute Worksheet 6-2, "Preparing to Make a Model of the Human Skeleton." Explain that the completed worksheet will serve as a "guide" for putting the model together and for sketching the skeletal system on the model.

Go over the directions with students. Direct them to:

1. Answer the questions according to the drawing.
2. Complete the drawing by sketching in bones on the blank side to match bones in the drawing.
3. Go over the answers together.

Answers to Worksheet, "Preparing to Make a Model of the Human Skeleton."

1. 1 bone in the upper arm.
2. 2 bones in the lower arm.
3. 2 bones in the thumb.
4. 3 bones in each finger.
5. 1 bone in the upper leg.
6. 2 bones in the lower leg.
7. 2 bones in the big toe.

Name _____ **Date** _____

Preparing to Make a Model of the Human Skeleton

You are going to make a model of the human skeleton. To prepare to make this model:

1. Answer the following questions according to the drawing.
2. Complete the drawing. Fill in the bones on the blank side by copying what you see in the drawing.

Answer: According to the drawing, how many bones are in:

1. The upper arm? _____

2. The lower arm? _____

3. The thumb? _____

4. Each of the other fingers? _____

5. The upper leg? _____

6. The lower leg? _____

7. The big toe? _____

8. Each of the other toes? _____

9. The rib cage and attached to the

   front of the rib cage? _____

10. The rib cage but not attached to the

    front of the rib cage? _____

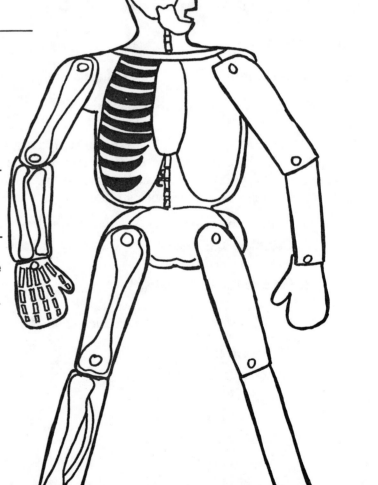

8. 3 bones in each of the other toes.

9. 10 ribs are attached to the front of the rib cage.

10. 2 ribs are attached to the backbone but not the front of the rib cage. They are called floating ribs.

OTHER POINTS FOR DISCUSSION

On a skeleton:

1. There is no smile; there are no lips.

2. There are no eyes.

3. There is no nose. The very top of the nose is bone; the rest is cartilage.

Lesson 6. Students Trace, Cut, and Store the Pieces for the Human Body Model

OBJECTIVES

Students will:

1. Follow multi-step directions to make a 13-piece puzzle of the human body model.

2. Refer to completed Student Worksheet 6–2, "Preparing to Make a Model of the Human Skeleton," to put punch holes in the pattern pieces and assemble the model.

MATERIALS

Completed Student Worksheets 6–2, "Preparing to Make a Model of the Human Skeleton"

At the supply cart:

One piece of oaktag (7″ by 3′) for each student

On each tray:

One envelope containing the 13 pattern pieces for the silhouette

A pair of scissors for each student (Closed and pointed down in a container)

One hole puncher for every two students

Four paper clips for every student

INSTRUCTIONAL PROCEDURES

Following Directions/Hands-on Activity

List the directions on the board and review them orally:

1. Trace and cut out the 13 pieces.
2. Return the pattern pieces to the envelope.
3. Write your name on the skull and one foot.
4. Punch out holes for joints as indicated by circles on the worksheet.
5. Stack the pieces so that it is easy to read the names. Use four paper clips to hold all of the pieces together.
6. Store scissors closed and pointed end down in the container.

If you finish early:

1. Put your stack of pieces away.
2. Help a friend.
3. Practice drawing the skeleton.

If you don't finish:

1. Use any free time that you have during the school day.
2. Finish in the next science class.

Teaching Tip

It is tempting to try to save time and have the students trace, cut, and put the model together in one class period. Only those students who have excellent small muscle movement and are good at following multi-step directions will be able to do this in one class period. Thus, it is recommended that you plan at least two class periods to make a model of the human skeleton from start to finish.

Lesson 7. Putting the Pieces Together

OBJECTIVES

Students will:

1. Follow multi-step directions.
2. Put the 13 pieces together to make a silhouette of the human body.
3. Sketch the major bones, in pencil, on the silhouette.
4. Outline the major bones in black marker.

MATERIALS

Completed student copies of Student Worksheet 6–2, "Preparing to Make a Model of the Human Skeleton"

On the teacher's desk, black felt tip markers (one per student)

On each tray:

Scissors (closed and pointed down in a container)

Hole puncher

Brass head paper fasteners (12 per student)

INSTRUCTIONAL PROCEDURES

List directions on the board. Go over them orally:

1. Finish cutting out and punching holes in the tracings, if necessary.
2. Lay out the pieces, according to the worksheet.
3. Fasten the pieces together, according to the worksheet, using the 12 fasteners.
4. Use a *pencil* to sketch the bones on one side of the model.
5. Have the teacher check the pencil sketch of the skeletal system.
6. Go over the sketch with a marker.

If you finish early:

1. Put your "skeleton" in the storage area.
2. Help a friend.
3. Practice drawing the brain and the nervous system which will be the next system studied. Use pictures in books or posters for reference.

Teaching Tips

1. To avoid some awkward statements, decide how you are going to refer to the model of the human body. My students call it a "skeleton." "Silhouette" would work. This will eliminate such statements as, "Put your body on the shelf." "Where is your model?"

2. Do not give a student a marker to outline the skeleton until you have checked the sketch. It is easy to make a correction on the pencil sketch.

3. Do not expect anyone to draw all 206 bones on the skeleton. The following requirements are reasonable.

- On the *skull:*

No smile, eyes or lips.

Some lines to show nonmovable joints.

- On the *rib cage:*

  Ten ribs attached to the rib cage.

  Two ribs attached to the backbone.

- On each *leg:*

  Upper leg . . . one long bone.

  Lower leg . . . two long bones.

- On each *arm:*

  Upper arm . . . one long bone.

  Lower arm . . . two long bones.

- On each *hand:*

  Four fingers . . . three short bones.

  One thumb . . . two short bones.

  Palm . . . one short bone from the wrist to the thumb; one bone from the wrist to each of the four fingers.

- On each *wrist* . . . nine small bones forming a gliding joint.

- On each *foot:*

  Four toes . . . three short bones

  One big toe . . . two short bones

  Heel . . . one bone

  From the toes to the heel . . . five bones

Lesson 8. Introducing the Nervous System and Voluntary Muscle Movements

OBJECTIVES

Students will:

1. State that the nervous system controls and regulates the body.
2. State that the brain is the control center of the nervous system.
3. Identify at least two voluntary actions, such as writing, walking, or swimming.

MATERIALS

Teacher-made model of the human body

Poster of the nervous system

Student Worksheet 6-3, "The Nervous System" (Younger students will need more assistance.)

Name _____ Date _____

The Nervous System

Directions: Read the following information about the nervous system.

     The nervous system controls the body and all of the individual systems. The brain is the control center for the nervous system. Nerves attached to the brain and spinal cord branch out throughout the body getting smaller and smaller. The nerve endings pick up sensations and send messages to the brain. The brain then sends a message to appropriate muscles. The brain is also responsible for all mental activity: thinking, learning, and problem solving.

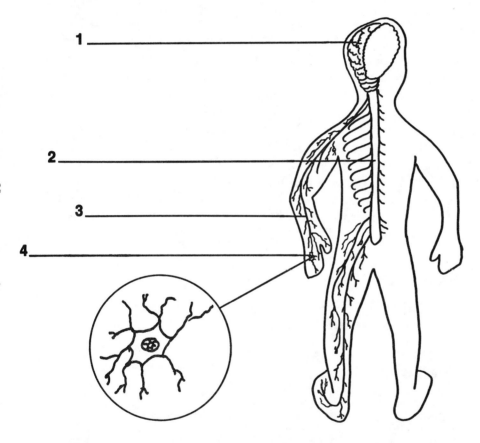

Directions: Label the diagram by matching the numbers and filling in the blanks. Complete the drawing. Reread the paragraph.

1. Brain

2. Spinal cord

3. Nerve fiber

4. Nerve cell

INSTRUCTIONAL PROCEDURES

Student Activity

Direct students to write their name and then put the pencil down. Say:

1. Did you have to think about how to use your fingers to pick up the pencil? (No)
2. Think about when you first learned how to print and write.

Explanation

Writing is a voluntary action. All movement is controlled by the nervous system, with the brain as the control center. Writing, swimming, riding a bike are all examples of voluntary actions which we learn. We get better with practice. Eventually we can do them without thinking about them. The nervous system controls these voluntary movements.

Distribute the Student Worksheet 6-3, "The Nervous System." Review the directions. Explain that it would be impossible to draw all of the nerves on the outline, but they should show the spinal nerves come off the spinal cord and branch out, smaller and smaller, to all parts of the body.

When students are finished, discuss the answers. Direct students to correct any answers that they missed so that they will be able to use this as a study sheet.

Answers to Worksheet, "The Nervous System"

1. Brain   3. Nerve fiber
2. Spinal cord   4. Nerve cell

Lesson 9. The Reflex Action

(Grades 5–8; Can be adapted to a minilecture for younger students.)

OBJECTIVES

Students will:

1. Define a reflex action as an involuntary action resulting when a message is carried by a nerve to the spinal cord to a muscle which causes movement.
2. Explain that reflex actions protect the body.
3. Distinguish between voluntary and involuntary actions.

MATERIALS

> Poster of the nervous system
> Completed Student Worksheet 6–3, "The Nervous System"
> Student Worksheet 6–4, "Ouch"
> Book (or something with which to make a loud, unexpected noise)

INSTRUCTIONAL PROCEDURES

Hold up the completed model of the human body. Show the side with the nervous system. In a *soothing voice,* state that the nervous system controls the body and that the brain is the control center. As you do so, make a loud, surprising sound by slamming a large book on the floor. Then say, "Raise your hand if you:

- Jumped.
- Blinked your eyes.
- Can feel your heart beating faster.
- Took in a deep breath.
- Yelled.
- Put your hands to your face.

Explanation

Your body is "programmed" to jump away from potential harm without you having to think about it. If you place your hand on a hot burner you don't wonder, "Should I move my hand?"

The nerve endings in your fingers and hand pick up the messages of heat and pain and send them to your spinal cord which sends a message to your arm muscle. The arm muscle pulls your arm away from the source of pain and potential damage. The message also goes to the brain. If you had to think about what to do and then do it, you would have a much more serious burn. This movement is called an involuntary action or a reflex. Reflexes help protect us.

Student Activity

Team good readers with poor readers. Distribute, Student Worksheet 6–4, "Ouch!" Set the purposes for reading:

1. Read the paragraph first.
2. Use the diagram to locate the names of the parts of the nervous system and then fill in the blanks.

Allow about 20 minutes for students to complete the worksheet. Discuss the answers and direct students to make any corrections that are necessary.

Name _____ **Date** _____

Ouch!

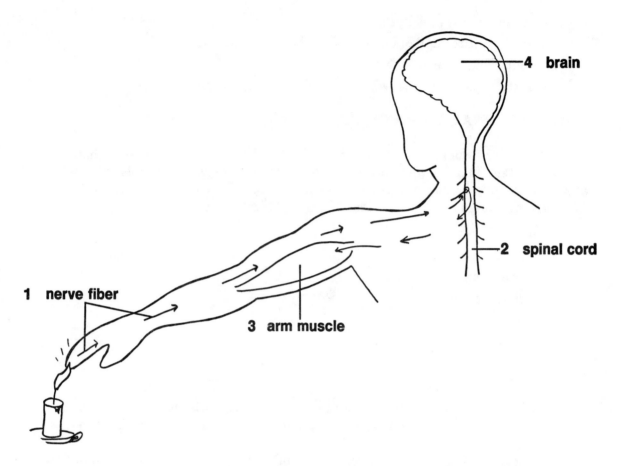

Directions: Use the diagram to answer the following questions. The numbers and letters serve as clues.

When We Say Ouch

A person accidentally touches the flame when reaching for a lighted candle. He pulls his hand back and yells. What happens inside that person?

1. The part of the body which carries the message of heat and pain is in the

   (1) __n__ ___ ___ ___ ___ ___   __f__ ___ ___ ___ ___.

2. It carries the message to the (2) __s__ ___ ___ ___ ___ ___   __c__ ___ ___ ___.

3. From there the message is sent two places. It is sent to the (3) __a__ ___ ___

   __m__ ___ ___ ___ ___ ___ which pulls the arm back.

4. The message continues to the (4) __b__ ___ ___ ___ ___ which receives and uses the information. Then the person usually yells and asks for some help.

Answers to Worksheet 6-4, "Ouch!"

1. Nerve fiber   3. Arm muscle
2. Spinal cord   4. Brain

Lesson 10. The Brain

For students in grades 5–8.

For younger students make a transparency or drawing on the board. Use the information for a minilecture.

OBJECTIVES

Students will:

1. State that the brain is the control center of the body.
2. Explain that the brain is made up of three parts, each of which performs a different function in controlling the body.
3. Identify the three parts of the brain as the cerebrum, cerebellum, and the medulla.

MATERIALS

Student Worksheet 6-5, "The Human Brain"
Model or poster of the brain

INSTRUCTIONAL PROCEDURES

Team poor readers with good readers. Distribute the Student Worksheet 6-5, "The Human Brain." Direct students to:

1. Fill in the blanks by reading the information and looking at the diagram.
2. Practice drawing the brain.

Allow about 25 minutes for answering the questions. Review the worksheet with the class.

Answers to Worksheet 6-5, "The Human Brain."

1. Three   4. Medulla
2. Cerebrum   5. Asleep
3. Cerebellum

Name _____   Date _____

The Human Brain

Directions: Use the diagram of the human brain to fill in the blanks.

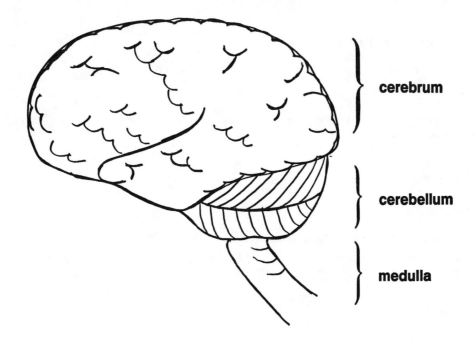

cerebrum

cerebellum

medulla

The human brain is made up of (1) ___ ___ ___ ___ _e_ parts. Each part controls

particular functions of the body.

The largest part of the brain is the (2) ___ ___ ___ ___ ___ ___ ___ _m_ . It controls

the conscious, rational thought, memory, intelligence, and all voluntary muscles, and receives

messages from the five senses. A smaller part, located below and behind the cerebrum is the

(3) ___ ___ ___ ___ ___ ___ ___ ___ ___ ___ _m_ . It controls the balance and

involuntary muscles.

The (4) ___ ___ ___ ___ ___ ___ _a_ is at the base of the skull, attached to the spinal

cord. It controls the involuntary muscles used in breathing, digesting food, pumping blood, and

other processes that must continue whether we are awake or (5) ___ ___ ___ ___ ___ _p_ .

POINTS FOR FURTHER DISCUSSION

1. The cerebrum is divided in half. The right side controls the voluntary muscles on the left side of the body, and the left side controls the voluntary muscles on the right side of the body.

2. Learning power can be increased by:

    a. organizing information
    b. using memory devices
    c. using practice devices such as outlines, drawing, and so forth

3. Injury to the nervous system can be serious. If we cut ourselves, we expect it to heal. If we break a bone, we expect it to mend in six to eight weeks. If we cut a nerve, we think that it will heal, but nerves are very complicated and usually do not regenerate. That means that injury to the brain, spinal cord, and nerves is usually permanent. However, the brain has many more cells than we use. Many times healthy or undamaged brain cells can be trained to take over the job of the damaged brain cells. Since damage to the brain and any part of the nervous system is dangerous and serious, everyone should develop safety awareness and safety habits for work and play.

4. Safety habits that students can develop:

    a. Don't take foolish or unnecessary chances.
    b. Wear protective clothing and gear for sports.
    c. Train for sports.
    d. Work on good posture.
    e. Learn the facts about the effects of drugs on the nervous system.

5. Drugs affect the nervous system. Drugs are chemical compounds that modify the body's natural chemistry of the cells. They are transported throughout the body by the blood system. If a person were going to have surgery, that person would most likely want a chemical to alter his body's perception of pain so that it cannot be felt.

For Your Information and Possible Discussion

Drugs can be classified as prescription drugs and nonprescription drugs. Prescription drugs are prescribed by a doctor. Just because a drug is prescribed by a doctor does not mean that it can be treated lightly or considered one hundred percent safe. Drugs can cause allergic reactions, harmful side effects, interfere with the development of a fetus, and sometimes be toxic. Care must be taken to keep all drugs out of reach of small children.

Nonprescription drugs can be classified as over-the-counter drugs and street or illegal drugs.

The oldest and most widely used drug is alcohol. It is a depressant. That means it slows down body processes such as breathing, heartbeat, and other muscle movements. Depending on the person and the amount of alcohol, the slowing down of the involuntary muscle movements can result in a minor problem or be as serious as death. It can interfere with judgment and result in an automobile accident.

There are a wide variety of drugs that affect the body differently. There are drugs that relieve pain, fight infection, influence the mind, stimulate the brain's control mechanisms, and depress or slow them down.

People who take illegal drugs or misuse prescription drugs and alcohol usually do so to feel better: a "pick up," a soothing effect or celebration. Remember a drug is a chemical. It enters the brain through the blood system and alters the controls. Hallucinogens such as LSD, produce sounds and sights that are not real. Stimulants such as caffeine, nicotine, and amphetamines, speed up the "controls." Depressants such as barbiturates and tranquilizers slow down the "controls." Drugs taken in large amounts or in combinations can result in serious health problems or death.

Teaching Tips

If your students express interest in asking for information about the effects of drugs, you may want to consider:

1. Attending some drug education or substance abuse education seminars.

2. Spending more than one day on this lesson. In later lessons, discuss the effects of some drugs on the systems that are being investigated.

3. Discussing the "Drug of the Day." Different drugs become popular at different times. LSD was popular in the 1960s. Cocaine became popular in the 1980s. Students see this in the news, in the neighborhood, and sometimes in their own home. They worry about the problems they observe. They don't need a sermon. They need facts. Some are relieved to learn that there are local support groups.

4. Providing students with guidance in recognizing what they can actually do something about and what they cannot. Guide them to developing personal choices and habits that promote good health and safety habits rather than concentrating on the pharmacological effects of drugs.

Lesson 11. Students Draw the Nervous System on the Model of the Human Body

OBJECTIVES

Students will draw the nervous system on the blank side of the human body model.

MATERIALS

Teacher-made model of the human body

Student models of the human body

Patterns of the human brain

Student Worksheets 6–3, "The Nervous System" and 6–5, "The Human Brain"

INSTRUCTIONAL PROCEDURES

Use the completed model of the human body to show that the nervous system is drawn on the blank side of the silhouette, not over the skeletal system.

Write student directions on the board:

1. Trace the pattern of the brain on the back of the skull.
2. Use a pencil to sketch the spinal cord.
3. Sketch spinal nerves coming off the spinal cord and branching out to all parts of the body.
4. Use the worksheets for reference.
5. Get the teacher to check the sketch.
6. Make any corrections that are necessary.
7. Go over the sketch with a marker.
8. Put the "silhouette" away.

If you finish early:

1. Help a friend.
2. Practice drawing the heart which is the major organ of the circulatory system, the next system to be studied.

Lesson 12. Introducing the Circulatory System

(Need to modify for Grades 3 and 4.)

OBJECTIVES

Students will:

1. State that the circulatory system delivers oxygen and food to all parts of the body.
2. State that the heart is a muscle which pumps the blood throughout the body.

MATERIALS

Completed model of the human body
Model or poster of the heart
Student Worksheet 6-6, "The Human Heart"

INSTRUCTIONAL PROCEDURES

Show the placement of the heart on the teacher-made model. Point out that the heart is located in the middle of the chest cavity and protected by the rib cage, and especially by the tough sternum running down the center of the chest. Ask, "Is the heart really shaped like a valentine?" (No)

"Is it the center of love?" (No)

Explanation

The heart is shaped like an upside down pear. It is not the center of love. It is a strong muscle which pumps the blood throughout the entire body every minute of your life.

The Chemistry of Blood

The blood carries food, oxygen, and chemicals to every cell. This provides the cells with nutrients and energy needed for playing, working, thinking, talking, and so forth. As the blood delivers nutrients and oxygen, it picks up carbon dioxide and other wastes. Some of the wastes are processed by the excretory system. The lungs give off the carbon dioxide and water vapor and pick up fresh oxygen.

Student Activity

Tell students that their heart is about the size of their own fist. Direct them to:

1. Make a fist.
2. Place it in the center of the chest.
3. Contract and relax the fist to get an understanding of how the heart muscles contract and relax.

When the muscles contract, they squeeze blood out of the "sending" chambers. When they relax, they permit blood to pour into the "receiving" chambers.

Place the students in reading teams. Distribute Student Worksheet 6-6, "The Human Heart." Explain the directions and allow about 20 minutes for students to answer the questions.

Discuss the information and answers with the students. Remind them to make any corrections that are necessary to make this a study sheet.

Name _____   Date _____

The Human Heart

The human heart is a muscle which pumps blood throughout the body. The blood carries food and oxygen to all parts of your body and picks up waste products.

The human heart is divided into four chambers. Each chamber has a special job to do:

- Chamber 1 (right auricle) receives the blood that has delivered food and oxygen and picked up waste products.
- Chamber 2 (right ventricle) contracts and sends the blood to the lungs to pick up fresh oxygen and drop off the waste.
- Chamber 3 (left auricle) receives the oxygen-rich blood from the lungs.
- Chamber 4 (left ventricle) contracts and sends the oxygen-rich blood to all parts of the body.

After the blood has nourished and cleansed the cells, it is ready to return to Chamber 1 and begin the trip again. The round trip takes about 20 seconds.

Using What You Have Read

Each of the four chambers in the heart has a name. <u>Find the name of each chamber and write it on the diagram.</u>

Most diagrams of the heart are colored red and blue. Red represents oxygen-rich blood and blue represents oxygen-poor blood. The left side of the heart receives the oxygen-rich blood which is bright red. <u>Color the left side of the heart, Chambers 3 and 4, bright red.</u>

The right side of the heart receives the oxygen-poor blood which is really dark red in color. <u>Color the right side of the heart, Chambers 1 and 2, blue.</u>

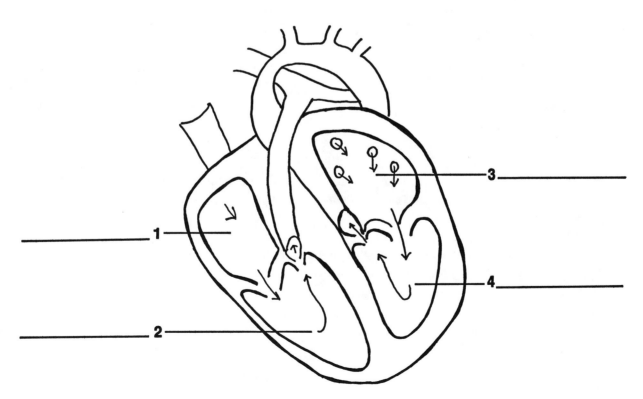

Answers to Worksheet 6-6, "The Human Heart"

1. Chamber 1 is named the right auricle.
2. Chamber 2 is the right ventricle.
3. Chamber 3 is the left auricle.
4. Chamber 4 is the left ventricle.

Teaching Tip

Introduce elementary students to the names of the four chambers but do not require them to memorize the names. Refer to them as the "four chambers" or the "sending" and "receiving" chambers.

Lesson 13. How Fast Does Your Heart Beat?

OBJECTIVES:

Students will:

1. Take pulse rates and record them on a table of organized data.
2. Make at least one statement of inference about the effect of exercise on the pulse rate.

MATERIALS

Classroom clock with a second hand.
Student Worksheet 6-7, "Pulse Rate"

INSTRUCTIONAL PROCEDURES

Distribute Student Worksheet 6-7, "Pulse Rate." Go over directions orally.

DISCUSSION

1. Pulse rate is affected by exercise. Generally speaking, exercise increases the pulse rate; rest decreases the pulse rate.

2. According to the American Medical Association (AMA), the heart needs a balance of exercise and rest. The heart is a muscle. Muscles need exercise; but they also need rest.

3. Sitting most of the day in school and most of the evening in front of the TV does not provide an adequate balance of exercise and rest.

4. When starting on a new exercise program, an individual should consult a physician and, in the case of children, their parents.

5. The AMA recommends a good night's sleep of at least eight hours and two rest periods a day.

Name ⎯⎯⎯⎯⎯⎯⎯⎯⎯⎯⎯⎯⎯⎯⎯⎯⎯⎯   **Date** ⎯⎯⎯⎯⎯⎯⎯⎯⎯⎯⎯⎯⎯⎯

Pulse Rate

Your heart pumps blood through elastic tubes called arteries. As it does so, the arteries stretch and resume regular size. This movement can be felt many places. It is called the pulse. To take your pulse:

1. Place your fingers on your neck, on either side of the windpipe where your neck joins your head. Gently push and feel the throbbing.

2. Count the number of throbs in one minute for your pulse rate.

Directions for the science activity: Take your pulse rate for each of the activities listed below, and record the results on the chart. Write at least one inference about the effect of activity on the pulse rate.

Pulse Rate

Activity	Number of Throbbings per Minute
While sitting	
While standing	
After jumping in place for 1 minute	
After resting for 1 minute	
After resting for 3 minutes	

Inferences About the Effects of Exercise and Rest on the Pulse Rate

⎯⎯⎯⎯⎯⎯⎯⎯⎯⎯⎯⎯⎯⎯⎯⎯⎯⎯⎯⎯⎯⎯⎯⎯⎯⎯⎯⎯⎯⎯⎯⎯⎯⎯⎯⎯⎯

⎯⎯⎯⎯⎯⎯⎯⎯⎯⎯⎯⎯⎯⎯⎯⎯⎯⎯⎯⎯⎯⎯⎯⎯⎯⎯⎯⎯⎯⎯⎯⎯⎯⎯⎯⎯⎯

Lesson 14. The Tubes That Carry Life

Note: This is an optional lesson for students in grades 6–8.

OBJECTIVES

Students will:

1. State that there are miles of blood vessels in the circulatory system.
2. Identify blood vessels as arteries, veins, and capillaries.
3. Explain that delivery of oxygen and food as well as discharge of carbon dioxide and other waste takes place in the capillaries.

MATERIALS

Poster of the circulatory system, colored red and blue
Information Sheet 6-A, "The Tubes of Life"
Student Worksheet 6-8, "The Tubes of the Circulatory System"
Red and blue crayons

INSTRUCTIONAL PROCEDURES

Team good readers with poor readers. Set the purposes for reading. Direct students to:

1. Work in the assigned teams.
2. Read the questions first.
3. Read the information looking for the answers.
4. Record the answers on the worksheet.
5. Color the arteries and capillaries branching off the arteries, red.
6. Color the veins and the capillaries branching into the veins, blue. Note the "coloring line" on the drawing.

Distribute the Information Sheet 6–A, "The Tubes of Life" and Student Worksheet 6–8, "The Tubes of the Circulatory System." Allow a class period for students to complete the worksheet. In the following class, discuss each answer, and have students make any corrections that are necessary.

Answers to Worksheet 6-8, "The Tubes of the Circulatory System"

1. Arteries, capillaries and veins.
2. a. Nutrients are delivered to every part of the body; and waste products are picked up for disposal.
   b. Carbon dioxide is exchanged for oxygen.

The Tubes of Life

Blood travels through a network of branching elastic tubes to reach every part of the body. To get from the heart to other organs and cells, the blood travels through 60,000 miles of blood vessels, the tubes of life.

There are three types of blood vessels: arteries, veins, and capillaries. It is in the main artery that the blood starts its long journey.

The blood is squeezed out of the left ventricle (Chamber 4) of the heart into the aorta, the main artery. The aorta branches into progressively smaller arteries. These arteries continue to branch throughout the body getting progressively smaller. They become so small that they can be seen only with a microscope. These microscopic tubes are called capillaries.

The walls of the capillaries are extremely thin. So thin, that body chemicals can pass through them. In the capillaries, blood delivers nutrition and oxygen to cells and picks up carbon dioxide and other waste.

The capillaries branch out, getting progressively larger and larger, forming the veins which carry the blood back to the heart and lungs for fresh oxygen and a new journey.

Blood flows in one direction: from the arteries to capillaries to veins.

Name _____   Date _____

The Tubes of the Circulatory System

Directions: Answer the following questions according to the information sheet, "The Tubes of Life." Color the arteries and the capillaries branching off the arteries, red. Color the veins and the capillaries branching into the veins, blue.

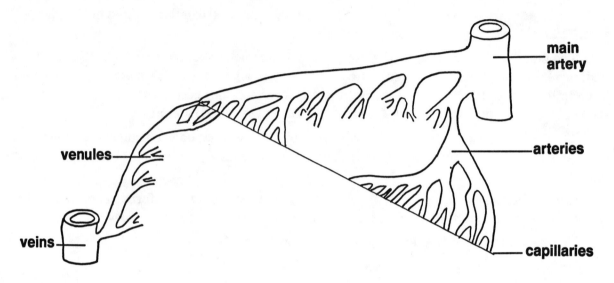

1. Names of the three types of "tubes of life."

   a. _____

   b. _____

   c. _____

2. Explain the two important functions that take place where the capillary meets the cell.

   a. _____

   b. _____

3. Name the tubes that carry oxygen-rich blood away from the heart throughout the entire body.

   _____

4. Name the tubes which are so thin that the blood passes through them to deliver food and oxygen

   and to pick up waste products. _____

5. Name the tubes that transport the blood carrying the waste products back to the heart.

   _____

3. Arteries.

4. Capillaries.

5. Veins.

FOR FURTHER DISCUSSION

Consider some findings and recommendations of the American Heart Association:

1. A low fat, low salt diet is recommended. Fat can accumulate in the vascular system, narrowing the passageway and causing the heart to work harder. Salt can also make the heart work harder and increase the blood pressure.

2. Nicotine, found in cigarettes, restricts arteries which also cause the heart to work harder and increases blood pressure.

3. Smoke from other people's cigarettes and other forms of air pollution make the heart work harder.

4. Stimulants such as caffeine found in coffee, tea, colas, and chocolate make the heart work harder.

5. Stimulants found in some medications and other drugs make the heart work harder. Some examples are amphetamine, benzedrine, and Speed.

6. Depressants slow the heart down. Some examples are tranquilizers and alcohol.

7. The heart is affected by what we eat, drink, and breathe. It is important to develop health-promoting habits.

8. A good night's sleep plus two short rest periods are recommended.

9. Exercise causes the heart to beat faster which exercises the heart, arteries, and veins, and increases the effective use of oxygen by the body.

What You Can Do to Help Yourself:

1. Eat a balanced diet.

2. Get plenty of rest and exercise.

3. Avoid the use of substances that are known to be harmful.

4. Strive for moderation. Those who eat, drink stimulants and depressants, smoke, worry, and argue to excess put stress on the heart.

5. Work on your own personal health and safety habits; don't nag your parents and friends. No one likes to be nagged.

Lesson 15. Students Put the Heart on the Model of the Human Body

OBJECTIVES

Students will follow multi-step directions to make a cutout of the heart and glue it on the model of the human body left of the center in the chest.

MATERIALS

Student models of the human body
On each tray:
  Patterns of the heart
  Unlined paper
  Scissors
  Red and blue crayons
  Glue and popsicle sticks

INSTRUCTIONAL PROCEDURES

Hands-on Activity

Write student directions on the board:

1. Trace the pattern of the heart.
2. Color the heart red; outline some veins blue.
3. Cut the tracing and glue it to the center of the chest.

If you finish early:

1. Help a friend.
2. Practice drawing the digestive system.

Lesson 16. Introducing the Digestive System

OBJECTIVES:

Students will:

1. State that the digestive system chemically changes food so that the body can use it for growth and energy.
2. Identify the major parts of the digestive system as the mouth, esophagus, stomach, and intestines.

3. Explain that the gall bladder aids in digestion.

4. Explain that human waste is a result of the digestive process.

MATERIALS

Poster of the digestive system
Teacher-made model of the human body
Reference material on the digestive system
Student Worksheet 6-9, "The Digestive System"

INSTRUCTIONAL PROCEDURES

Distribute Student Worksheet 6-9, "The Digestive System." Direct students to label the names of the seven parts of the digestive system by matching the numbers on the diagram with the numbered list of names on the worksheet.

Allow about 20 minutes for students to complete the worksheets. Identify the names of the parts of the digestive system. Direct students to make any corrections that are necessary. Explain and discuss how the digestive system works.

Answers to Student Worksheet 6-9, "The Digestive System."

1. Mouth
2. Esophagus
3. Stomach
4. Small intestine
5. Large intestine
6. Gall bladder
7. Liver

Explanation of the Digestive System

Every part of the body must receive nourishment. The body gets its nourishment from the food that we eat. But the food is too big and too complex to be of any use. The digestive system chemically changes the food into a form that can be used for growth and energy.

The chemical changes begin when we start chewing. As teeth grind food, they mix it with saliva which pours into the mouth from six salivary glands. Saliva softens food and chemically changes starches into sugars.

The tongue helps move the saliva softened food to the back of the mouth and muscles squeeze the food into the food tube called the esophagus. There the food is squeezed down into the stomach by muscles which surround the esophagus. Muscles, not gravity, move food through the entire digestive system. That is why astronauts can eat in the weightless environment of space.

While in the stomach, food is churned with the chemicals of the stomach: enzymes and dilute hydrochloric acid. Enzymes are chemicals that cause chemical reactions but are not affected by them. Hydrochloric acid is truly an acid. The

Name _____   Date _____

The Digestive System

Directions: Label the parts of the digestive system. Match the numbers in the diagram with the numbers next to the names of the parts of the digestive system.

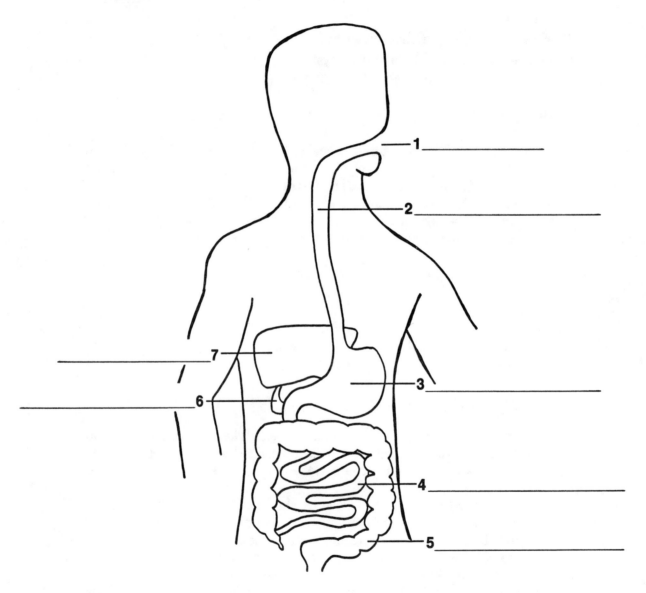

1. Mouth
2. Esophagus
3. Stomach
4. Small intestines
5. Large intestines
6. Gall bladder
7. Liver

stomach's main job is to chemically break down protein (foods from animals, nuts, beans, and peas) into a form which can travel through the circulatory system. It does not digest much fat or starch.

The stomach has a protective lining which protects itself against the acid. However, too much acid can irritate the lining of the stomach, and cause pain and ulcers. Antacids can neutralize excessive acids.

After two to six hours of churning in the stomach, the partially digested food is squeezed by muscles into the small intestines. The larger the meal, the longer the time required for digestion.

The gall bladder contributes fluids which aid in digestion. The liver stores sugars and keeps the bloodstream from being flooded by foodstuffs.

The food that is thoroughly digested is absorbed into blood vessels at different stages of the intestines. The digested food is eventually transported, by the blood, throughout the entire body supplying usable nutrition to every system, organ, and cell.

In the cell, these nutrition products chemically combine with oxygen to provide growth and energy.

The food that is not thoroughly digested, the waste, is squeezed into the large intestine. The waste products are mostly made up of plant products that cannot be digested. The body cannot use these waste products and eliminates them from the body through the rectum in the form of a bowel movement.

Teaching Tips

1. This may take more than one class period.

2. Direct students to either color each of the seven parts a different color or different shades of the same color in order to become more familiar with the parts of the digestive tract. The digestive system is usually colored pink, with the liver colored dark red.

Lesson 17. The Beginnings of Digestion: Starches to Sugars

OBJECTIVES

Students will:

1. State that digestion begins in the mouth.
2. Explain that saliva chemically changes starch to sugar.

MATERIALS

An unsalted soda cracker or small piece of bread for each student

INSTRUCTIONAL PROCEDURE

Go over all of the directions. Give each student a soda cracker.

Student Activity

Direct students to:

1. Taste a small piece of the cracker.
2. Write a description of the taste.
3. Slowly chew the rest of the cracker until there is a change in the taste.
4. Write a description of the new taste.

Explanation

Our teeth tear and grind the cracker. The salivary glands located in the mouth deliver saliva when the sense of sight, smell, or taste detect food.

The cracker is made of starch. It has a bland taste at first. As we chew it, our teeth mix the cracker with saliva. Saliva contains ptyalin which is a chemical that changes the starch in the cracker into sugar. Thus, as we chew the cracker a gradual sweetening taste develops in our mouth. Changing starch to sugar is the first step in digestion.

Lesson 18. Students Put the Digestive System on the Model of the Human Body

OBJECTIVES

Students will follow multi-step directions to make a paper model of the digestive system and glue it on the model of the human body.

MATERIALS

Student models of the human body
Completed Worksheet 6–9, "The Digestive System"
On each tray:
      Patterns of the digestive system
      Unlined paper
      Scissors
      Crayons
      Glue and popsicle sticks

INSTRUCTIONAL PROCEDURES

Hands-on Activity

Direct students to:

1. Trace and cut out the pattern for the digestive system.
2. Draw and color the digestive system on the tracing. Refer to the worksheet.
3. Glue the mouth and esophagus to the model.

If you finish early:

1. Help a friend.
2. Read about the different food groups and identify the Basic Four.

Lesson 19. What Does the Body Do with the Digested Food?

Note: This is an optional lesson for students in grades 6–8.

OBJECTIVES

Students will explain that the body uses digested food for three purposes:

- growth and repair
- energy
- regulation of body processes.

MATERIALS

Student Worksheet 6-10, "What Does the Body Do with Digested Food?"

INSTRUCTIONAL PROCEDURES

Team good readers with poor readers. Direct them to read the information on the worksheet, then identify the body's uses of digested food and be prepared to discuss their answers.

Distribute the worksheets. Allow a class period to complete the worksheet and one to review it.

Answers to Student Worksheet 6–10, "What Does the Body Do with Digested Food?"

1. Nutrients in digested food provide materials for three important tasks. They are:

Name _____   Date _____

What Does the Body Do with the Digested Food?

Digested foods called <u>nutrients</u> are delivered to all parts of the body by the blood. These nutrients:

1. Provide for the growth and repair of cells and tissues.
2. Produce energy.
3. Regulate the body processes such as breathing.

There are seven nutrients: proteins, sugars, starches, fats, vitamins, minerals, and water. Each nutrient helps the body perform one of the three important jobs listed above. No <u>one</u> nutrient can provide materials to accomplish all three tasks. That is why we need a balanced diet.

Proteins provide the materials that are needed for cell growth and repair. Extra protein can provide energy.

Sugars, starches, and fats provide energy. They combine with oxygen in the cells to produce heat energy.

Vitamins, minerals, and water provide the chemicals needed to help the body function. The two best known minerals are calcium and phosphorus. They harden teeth and bones. Water is essential to life. It is the major part of the blood and the body.

Directions: Answer the following questions based on the information provided.

1. Nutrients in digested foods provide materials for three important tasks. The three tasks are:

   a. _____

   b. _____

   c. _____

2. Foods can be grouped into seven nutrients. Name the seven nutrients.

   a. _____

   b. _____

   c. _____

   d. _____

   e. _____

   f. _____

   g. _____

3. Explain why a balanced diet is necessary.

   _____

   _____

a. Growth and repair of body cells and tissue.

b. Production of energy.

c. Regulation of body.

2. Food can be grouped into seven nutrients. Name them.

a. Proteins   e. Vitamins

b. Starches   f. Minerals

c. Sugars   g. Water

d. Fats

3. Explain why a balanced diet is necessary.

No one food or type of nutrient can provide all the materials that the body needs to function. A balanced diet is needed to provide all of the ingredients for growth and repair, energy, and regulation of systems and processes.

Lesson 20. Identifying and Balancing the Basic Four Plus One

Note: This is an optional lesson for students in grades 6–8.

OBJECTIVES

Students will:

1. Identify the four basic food groups as dairy, protein, vegetable and fruit, and bread and cereal.

2. State that water is essential for life and that drinking eight glasses a day is recommended.

MATERIALS

Information Sheet 6-B, "Balancing the Basic Four Plus One"
Student Worksheet 6-11, "Identifying the Basic Four"
Resource materials on foods

INSTRUCTIONAL PROCEDURES

Day 1

Distribute Information Sheet 6–B, "Balancing the Basic Four Plus One." Review it with students by asking questions that they can answer from the information on the chart. For example:

1. What are the four food groups?

2. What is meant by "plus one?"

Name _____   Date _____

Balancing the Basic Four Plus One

No one food can provide all the nutrients that are needed for growth, repair, energy, and regulation of systems. A combination of the seven nutrients (protein, starches, sugars, fats, vitamins, minerals, and water) is needed. To help us plan and eat balanced meals, nutritionists who are experts in foods have divided all foods into four groups and have identified the number of servings per group that are recommended in one day to make a healthy diet. The following chart shows the four food groups and the recommended daily servings for each group, plus water, the fluid that is essential for life. Drinking eight glasses of water a day is recommended.

The Basic Four

The Dairy Group	Protein Group
Milk **Cheese** **Ice Cream** They supply minerals that build strong teeth and bones. Are usually fortified with vitamins. Recommended daily servings: Children . . . 4 or more cups Adults . . . 2 cups	Meat Fish Eggs Peanut butter Nuts Dried peas and beans They supply materials that are necessary for growth and repair of cells and fats for energy. Recommended daily servings: Everyone . . . 3 to 6 ounces
Fruit and Vegetable Group	Bread and Cereal Group
Fresh fruits and vegetables Raw or steamed preferred The longer they are cooked, the less their nutritional value. They supply sugars and starches for energy; along with vitamins and minerals. Everyone . . . 4 or more	Bread Cereal Rice Whole grain, enriched with vitamins. Not sugar coated. They supply sugars and starches for energy and frequently have added vitamins and minerals. Children . . . 4 or more Adults . . . 4

Plus water. Eight glasses are recommended.

Name _____ Date _____

Identifying the Basic Four

Directions: Use the information provided in, "Balancing the Basic Four Plus One," to identify sources of needed nutrients.

1. Four daily servings of dairy products are recommended. Name four dairy products.

    1. _____

    2. _____

    3. _____

    4. _____

2. Three servings from the protein group are recommended. Name three sources of protein.

    1. _____

    2. _____

    3. _____

3. Four or more servings of fruits and vegetables are recommended. Name five fruits and five vegetables.

   Fruits: Vegetables:

    1. _____   1. _____

    2. _____   2. _____

    3. _____   3. _____

    4. _____   4. _____

    5. _____   5. _____

4. Four or more servings from the bread and cereal group are recommended. Name four sources of bread and cereal.

    1. _____

    2. _____

    3. _____

    4. _____

5. How many glasses of water a day are recommended?

    _____

3. Why do you think water is so important?

4. How many servings of dairy products are recommended in one day?

5. How many servings of protein are recommended?

6. How many servings of fruits and vegetables are recommended?

7. How many servings of breads and cereals are recommended?

8. What nutrients do the dairy products supply?

9. What nutrients do proteins supply?

10. What nutrients do fruits and vegetables supply?

11. What nutrients do breads and cereals supply?

Day 2

Distribute Student Worksheet 6-11, "Identifying the Basic Four." Direct students to use the information from Information Sheet 6–B, "Balancing The Basic Four Plus One" to answer the questions. Review the answers.

Answers to Worksheet 6-11, "Identifying the Basic Four"

1. Name four dairy products. Possible answers:

milk	milk shakes
cottage cheese	ice cream
hard cheese	yogurt

2. Name three sources of protein. Possible answers:

meat	fish
eggs	shellfish
nuts	peanut butter
poultry	dried beans and peas

3. Name five fruits and five vegetables. Possible answers:

Fruits:	Vegetables:
apples	lettuce
pears	spinach
bananas	corn
oranges	potatoes
grapes	peas
grapefruit	broccoli
tangerines	cauliflower

4. Name four sources of bread or cereals. Possible answers:

Breads:	Cereals:
pumpernickel	Shredded Wheat
rye	All-Bran
whole wheat	Total
enriched white	Cereals that are not sugar coated

5. Eight glasses of water a day are recommended.

"If You Like" Assignments

1. Make and illustrate posters of the Basic Four.
2. Make posters showing easily obtainable sources for each nutrient.
3. Keep a record of how many servings of each food group that you have in a day.

Lesson 21. Introducing the Respiratory System

OBJECTIVES

Students will:

1. Name the breathing system—the respiratory system.
2. Identify the major parts of the respiratory system as the nose, trachea, bronchial tubes, lungs, and air sacs or alveoli.
3. Explain that the respiratory system puts oxygen in the blood and takes carbon dioxide and water vapor out of the blood.

MATERIALS

Teacher-made model of the human body

Poster of the respiratory system

Student Worksheet 6-12, "The Respiratory System"

INSTRUCTIONAL PROCEDURES

Show the students the location of the respiratory system on the cut-and-paste model of the human body. Ask students what the lungs do. Most know that the lungs are responsible for breathing, but few understand how gaseous oxygen gets from the air into the body fluids.

Name _____   **Date** _____

The Respiratory System

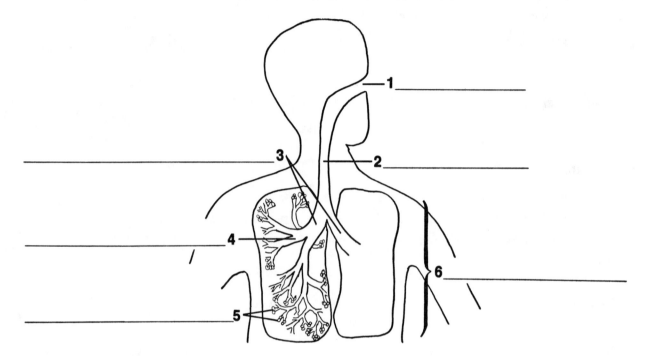

© 1989 by The Center for Applied Research in Education

Directions: Use the information and names in the paragraphs to label the parts on the diagram.

The Respiratory System

The respiratory system is the breathing system. Its major jobs are to exchange carbon dioxide for oxygen and get rid of water vapor.

The air takes a long journey through the many tubes of the respiratory system, which are moist with mucus and lined with small, hair-like cilia. The mucus and cilia help filter out air pollutants and bacteria.

The air enters the (1) nose. From there, the air travels down the windpipe or (2) trachea. When trapped particles irritate the nose and windpipe, they produce a sneeze or a cough to rid the system of the irritants.

The windpipe branches into two tubes called the (3) bronchial tubes. The bronchial tubes branch out into progressively smaller and smaller branches called (4) bronchioles. The very small bronchioles end in clusters of small, balloon-like air sacs called (5) alveoli.

Air travels through all of these tubes until it reaches the millions of elastic air sacs.

As we breathe in, the air sacs balloon up with the air causing the (6) lungs and chest to fill out or expand. As we breathe out, the air sacs deflate, causing the chest to contract.

In the air sacs, the oxygen prepares for the oxygen-carbon dioxide exchange.

In grades 5–8, team good readers with poor readers. Distribute Student Worksheet 6-12, "The Respiratory System." Inform students that all of the parts of the respiratory system can be labeled after reading the information on the worksheet. Allow 30 minutes for students to label the parts of the respiratory system. Go over the answers together and direct students to make any corrections that are necessary.

In grades 3 and 4, work through the sheet with the class.

Answers to Student Worksheet 6-12, "The Respiratory System"

1. Nose
2. Trachea or windpipe
3. Bronchial tubes
4. Bronchioles
5. Air sacs or alveoli
6. Lungs

FOR FURTHER DISCUSSION

1. Why is mouth breathing considered a bad habit? (The mouth does not have small hairs and cannot clean the air as effectively as the nose can.)
2. The pathways for air are lined with mucus and cilia which filter out some pollutants and bacteria. Thus the lungs can clean themselves when they are not overloaded with pollutants or bacteria. An overload can result in malfunction and/or disease.

Lesson 22. The Oxygen Exchange

Note: This is an optional lesson for students in grades 6–8.

OBJECTIVES

Students will:

1. State that air enters the air sac.
2. State that air sacs are surrounded by capillaries which carry blood.
3. Explain that the oxygen dissolves in the moist coating which lines the air sac.
4. Explain that the dissolved oxygen passes from air sacs into the blood in the capillaries where it is exchanged for carbon dioxide.
5. State that the oxygen is transported by blood throughout the body.
6. State that the carbon dioxide and water vapor are exhaled out of the body.

MATERIALS

Student Worksheet 6-13, "The Heart and the Lungs Work Together"

Name _____   **Date** _____

The Heart and the Lungs Work Together

   Air journeys from the nose, down the trachea, through the many branches of the bronchial tubes, through the smaller bronchioles, into the clusters of balloon-like air sacs.

   In the air sacs, oxygen dissolves in the moisture that lines the air sacs. The dissolved oxygen passes through the thin wall of the air sacs, through the thin wall of the surrounding capillaries into the blood. There, hemoglobin, a chemical in the blood, exchanges carbon dioxide for oxygen. The oxygen-rich blood travels through the circulatory system to the heart which pumps it to all parts of the body.

   Carbon dioxide and water vapor in the capillary pass through the thin walls into the air sac. There they travel to the bronchioles, through the bronchial tubes, up the trachea, and are exhaled through the nose or mouth.

   Place your hand a few inches from your mouth. Breathe out onto your hand. Can you feel the water vapor collect on your hand as you exhale?

Directions: Based on the diagram and the reading, fill in the blanks.

1. Air travels through the many branching tubes in the lungs, ending its journey in the

    __a__ ___ ___   __s__ ___ ___ ___.

2. Air sacs are surrounded by __c__ ___ ___ ___ ___ ___ ___ ___ ___ ___.

3. Dissolved oxygen passes through the wall of the air sac and capillary and is exchanged for

    __c__ ___ ___ ___ ___ ___   __d__ ___ ___ ___ ___ ___ ___.

4. The waste products, carbon dioxide and water vapor, pass through the capillary and enter tubes

    of the respiratory system and are __e__ ___ ___ ___ ___ ___ ___ out of the body.

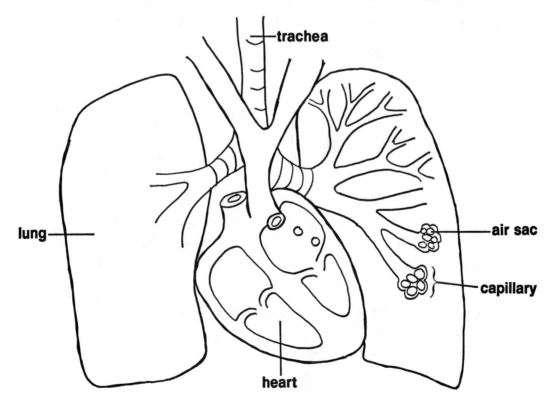

INSTRUCTIONAL PROCEDURES

Place the students in reading teams. Direct them to:

1. Read the information and answer the questions.
2. Sketch in the missing tubes of the respiratory system on the left lung to match the tubes on the right lung.

Note: You may want to explain that the left lung is slightly smaller than the right to make room for the heart.

Answers to Student Worksheet 6-13, "The Heart and the Lungs Work Together."

1. air sacs   3. carbon dioxide
2. capillaries   4. exhaled

Lesson 23. The Effects of Tobacco Smoke on the Respiratory System

OBJECTIVES

Students will:

1. Explain that tobacco smoke carries many harmful chemicals which travel throughout the branching tubes of the respiratory system.
2. Identify three of these chemicals as nicotine, tars, and carbon monoxide which can cause irritation and disease.
3. Explain that carbon monoxide is absorbed into the blood stream, reducing the amount of oxygen available to the body.

MATERIALS

Completed Student Worksheets 6–12, "The Respiratory System" and 6–13, "The Heart and the Lungs Work Together"

Student Worksheet 6-14, "Effects of Tobacco Smoke on the Lungs"

INSTRUCTIONAL PROCEDURES

Place students in reading teams. To set the purposes and stimulate interest in the reading ask the following questions. Listen to student responses without comment.

1. Have you noticed the warnings on cigarette packs or in cigarette ads?
2. Have you heard that smoke from other peoples' cigarettes can cause health problems?
3. Where do you think the tobacco smoke goes?

Name _____ **Date** _____

The Effects of Tobacco Smoke on the Lungs

Have you wondered what smoking does and what happens if a person quits?

Tobacco smoke contains harmful chemicals which travel throughout the tubes of the respiratory system, enter the air sacs, and disrupt the oxygen-carbon dioxide exchange.

Three harmful chemicals in smoke are: tars, nicotine, and carbon monoxide.

Tars irritate and dirty the lungs. They can cause cancer.

Nicotine stuns the hair-like cilia and prevents them from "sweeping" out the lungs. It also narrows the blood vessels causing the heart to work harder. In addition, it upsets the stomach.

Carbon monoxide enters the air sacs. The blood picks up the carbon monoxide in place of oxygen and exchanges it for carbon dioxide.

The heart then pumps oxygen-poor instead of oxygen-rich blood through the body. This causes the heart to work harder.

Smoke from other people's cigarettes also irritates the respiratory system.

Years of smoking can cause serious health problems. Air sacs lose the ability to inflate and deflate like balloons. Some blow up. This can cripple a lung and cause the disease emphysema. Smoking can also cause chronic bronchitis, lung cancer, and heart attacks.

When a smoker quits, the lungs begin to clean themselves. The cilia begin "sweeping" again. The mucus picks up and expels dirt and bacteria. The oxygen-carbon dioxide exchange is restored and the body gets the oxygen that it needs. The heart and lungs can resume their normal functions, if they have not been damaged.

Directions: Fill in the blanks. Be prepared to discuss your answers.

1. Tars can cause _____.

2. Nicotine stuns the _____.

3. Nicotine causes the blood vessels to become narrow, making the heart work _____.

4. Carbon monoxide interferes with the exchange of _____ for carbon dioxide.

5. When a smoker quits, the lungs begin to _____.

Direct the students to read and answer the questions on the worksheet and to be prepared to discuss the answers.

Answers to Student Worksheet 6–14, "Effects of Tobacco Smoke on the Lungs"

1. Cancer   4. Oxygen
2. Cilia   5. Function or work better
3. Harder

FOR YOUR INFORMATION AND POSSIBLE DISCUSSION

According to the American Lung Association, marijuana smoke contains 12 times the amount of tars and 20 times the amount of carbon monoxide as tobacco smoke.

Chewing tobacco can cause cancer of the mouth and jaw.

Potential Resource

For additional information, including booklets and tapes suitable for classroom use, contact your local chapter of the American Cancer Society and/or The American Lung Association.

Lesson 24. Unequal Air Pressure Causes Breathing

OBJECTIVES

Students will:

1. State that when the diaphragm pulls downward, the lungs pull air in and when the diaphragm pushes up, the lungs push air out.
2. Explain that breathing is caused by a difference of air pressure caused by the movement of the diaphragm and rib muscles.

MATERIALS

Top section of a plastic one-liter soda bottle
Glass straw
One-hole cork, stopper, or clay
Balloon
Two strong rubber bands
Rubber sheet or tough balloon cut open to make a rubber sheet

BEFORE CLASS

Build a Model of a Lung. See Figure 6-2.

1. Cut off and discard the bottom section of a plastic one-liter soda bottle.
2. Place the one-hole stopper in the mouth of the soda bottle.
3. Use a rubber band to secure one balloon to one end of the glass straw.
4. Insert the other end of the glass straw through the hole in the stopper.
5. Use a rubber band to secure the rubber sheeting to the opened end of the soda bottle.

Test Your Model Before Using It in Class

1. Pull down on the rubber sheet. The balloon inflates.
2. Push up on the rubber sheet. The balloon deflates.
3. If there are any leaks, seal with wax or clay.

INSTRUCTIONAL PROCEDURES

Demonstration

Identify the balloon in the bottle as a model of one lung inside the chest cavity. Ask what different parts of the model represent:

1. The glass straw? (The air passages in the respiratory system.)
2. The balloon? (A lung or an air sac)
3. The balloon stretched across the bottom of the bottle? (The diaphragm)

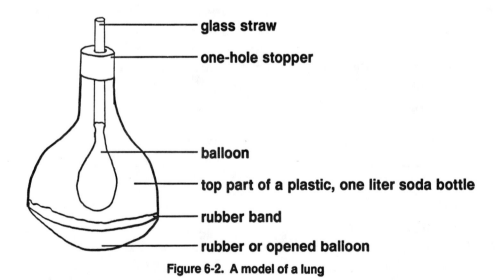

glass straw

one-hole stopper

balloon

top part of a plastic, one liter soda bottle

rubber band

rubber or opened balloon

Figure 6-2. A model of a lung

Introduce the new term "diaphragm" as the muscle which pulls downward and pushes upward on the lungs, causing some interesting things to happen.

Direct the students to watch the model as you pull downward on the rubber sheet. Ask for observations. (The balloon gets bigger.)

Ask for predictions if you push up on the rubber sheet. Listen to their predictions. Push up on the balloon. Ask for observations. (The balloon gets smaller. No more air goes in the balloon.)

Invite students to repeat the demonstrations.

Explanation

When the rubber sheet is pulled down, more room is made in the bottle. Increasing the room decreases the air pressure in the bottle, and the outside air pressure forces air into the balloon.

The lungs work in a similar manner. The diaphragm pulls down, increasing the chest cavity area, and reducing the air pressure in the lungs. The outside air pressure then forces air into the respiratory tubes and lungs and evens out the air pressure.

When the rubber sheet is pushed upward, less room is created in the bottle, causing greater air pressure inside the bottle. The air in the balloon is forced out.

When the diaphragm pushes upward, it makes less space in the chest cavity and more air pressure in the lungs. This forces the air out of the lungs.

POINTS FOR DISCUSSION

1. We can control our breathing for a while.
2. We do not need to think about breathing. It is controlled by the brain which sends messages to the muscle called the diaphragm. Breathing is an involuntary action.

Lesson 25. Students Put the Respiratory System on Their Models of the Human Body

OBJECTIVES

Students will follow multi-step directions to make a paper model of the respiratory system and place it on the model so that the mouth and part of the trachea are glued to the model. The lungs can be lifted to view other parts.

MATERIALS

Student models of the human body
Worksheets with drawings of the respiratory system

On each tray:

    Patterns for the respiratory system

    Unlined paper

    Scissors

    Glue and popsicle sticks

    Red and blue crayons

INSTRUCTIONAL PROCEDURES

Direct students to:

1. Trace the pattern of the respiratory system onto the unlined paper.
2. Sketch in the rings of cartilage on the trachea, the bronchial tubes, bronchioles, and air sacs.
3. Use red and blue crayons to make lines around some of the air sacs to indicate that the oxygen exchange takes place between the air sacs and the capillaries. (Grades 6–8)
4. Color the respiratory system pink or light red.

Teaching Tip

Have students put all of the worksheets together to make a booklet about the human body. Add articles from newspapers and magazines.

Lesson 26. Think About What You Have Learned About the Human Body

OBJECTIVES

Students will list at least four habits that would be desirable and possible to develop.

MATERIALS

All worksheets and posters on the human body

Paper and pencil

INSTRUCTIONAL PROCEDURES

Congratulate students on completing their models of the human body. Tell them that by working on this long-term project about the human body, they now possess important information that can be used to develop healthy habits and make important decisions.

Direct students to list four habits they could develop that would be healthy habits worth starting and keeping.

"If You Like" Assignment

Make an appointment with any adult in your family to show and explain the systems on your model of the human body. Since the explanations of the five systems will take some time, ask to do this when the adult has time to listen. Then write a note to me explaining what you have done and have the adult sign it; or have the adult write and sign the note.

Teaching Tips

1. Allow students or adults to write the notes. This allows the one who is interested in writing to do so.

The notes from my students' parents and grandparents have been fun and exciting to read. Parents are surprised to learn that their child is capable of learning so much about the body. Parents who are in the health professions have responded enthusiastically about the accuracy of the cut-and-paste model and student explanations.

2. During the first year that you teach this unit, don't try to teach everything in this section. Concentrate on:

   a. Helping each student complete a cut-and-paste model of the human body.
   b. Teaching one lesson about each system.
   c. Getting your own supplies organized.
   d. Developing a system for dispensing, collecting, and storing partially completed student-made models.
   e. Understanding the parts and functions of the five systems discussed in this section.

PERSONAL PERSPECTIVE

The first year that I taught my third graders how to make this model of the human body, I had a few patterns, some art supplies, a set of health texts, a modest bone collection, and a model of the heart. It took us eight weeks to complete our cut-and-paste models of the human body. The students and parents were proud; and so was I.

Making the Most of Your Magnets

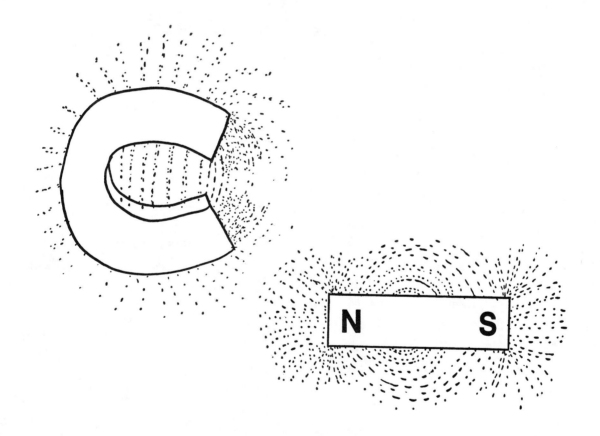

Personal Perspective

A pair of bar magnets can be a valuable teaching tool. By having your students observe their interaction with each other and other items, you can provide interesting learning experiences with inferring, counting, applying simple math, analyzing data, and solving problems and making games. The following is a series of lessons from which you can choose to help you make the most of your magnets.

Magnets, Computers, and Watches

Magnets erase information on computers and disks. They also magnetize the gears in mechanical watches. Thus, it is recommended that you teach students to keep magnets away from computers, floppy disks, and watches.

Lesson 1. Identifying and Classifying Objects That Are Magnetic and Are Not Magnetic

OBJECTIVES

Students will:

1. Classify objects as magnetic and nonmagnetic based on test results.
2. Record results on a table of organized data.

MATERIALS

Student Worksheet 7-1, "Classifying Objects by Magnetism"

On each supply tray place one magnet and eight to ten magnetic and not magnetic items: Suggested items:

Magnetic	Not Magnetic
Paper clip	Clothes pin
Iron nail	Aluminum nail
Thumb tack	Aluminum foil
Iron washer	Wax paper
	Milk carton
	Plastic cube
	Piece of clay

Name _____   **Date** _____

Classifying Objects by Magnetism

Magnetic	Not Magnetic

INSTRUCTIONAL PROCEDURES

Hold up a tray of supplies. Ask, "How can we find out which of these items are magnetic and which are not magnetic?" (Touch each with a magnet. If it sticks, it's magnetic. If it doesn't, it's not magnetic.)

Lab Directions

1. Test each object with a magnet.
2. Record its name under magnetic or not magnetic on the worksheet.
3. Those who finish early, test personal objects such as graphite in pencils, jewelry, and spiral rings on notebooks.

DISCUSSION

Identify which objects are magnetic and which are not magnetic.

Teaching Tip

Repeat this lab using different objects. This provides practice for mastery and a review for everyone, including anyone who missed the lab.

Lesson 2. What Types of Metals Are Magnetic?

OBJECTIVES

Students will:

1. Conduct a test to determine what types of metals are magnetic.
2. Write a conclusion based on the data stating that of the material tested, iron and steel are magnetic.

MATERIALS

Student Worksheet 7-2, "What Types of Metals Are Magnetic?"

On each tray place one magnet and eight to ten items to test. Suggested items:

Magnetic	Not Magnetic
Steel bolt	Aluminum bolt
Steel nut	Aluminum nut
Steel screw	Brass screw
Iron nail	Glass marble
	Piece of plastic

Name _____   **Date** _____

What Types of Metals Are Magnetic?

Magnetic	Not Magnetic	Conclusion

ON THE BOARD

List the items and identify their contents. For example:

steel bolt   iron nail
glass marble   wooden stick

INSTRUCTIONAL PROCEDURES

Ask students to make inferences about the types of things that are magnetic. Some may say that all metals are magnetic. Some may say that some, but not all, metals are magnetic.

Lab Directions

Ask, "How can we check these inferences?" If someone has a good idea, have that person demonstrate it. If not, direct students to:

- Select an item.
- Look at the board to determine what it is made of.
- Test it with a magnet.
- List it by name and makeup, such as iron nail, on the worksheet in the magnetic or not magnetic column according to the test.
- Write a conclusion about the types of things that are magnetic based on the data.

DISCUSSION

After cleanup, have students discuss their findings and read their conclusions. Acceptable conclusions:

- Some metals are magnetic, nonmetals are not magnetic.
- Of the metals tested, iron and steel are magnetic.

Lesson 3. Discovering What Types of Materials Are Magnetic by Reading (Grades 5–8)

OBJECTIVES

Students will work in reading teams to discover information about magnetism by reading a given worksheet.

Name _____   Date _____

About Magnetism

Directions: Read the questions first, then read the paragraphs and answer the questions.

Things Magnetic

Scientists think that all atoms can act like magnets. Everything is made of atoms, yet only three metals are attracted to a magnet. They are iron, nickel, and cobalt. Why aren't more things magnetic?

Scientists believe that it has to do with the way atoms are lined up in a material. It is easiest to think of it as scrambled and unscrambled alignment. If the atoms are aligned so that they are unscrambled and pointing in one direction, as shown in Figure 1, the material is magnetic. If the atoms are aligned so that they are scrambled and pointing in different directions, as shown in Figure 2, the material is not magnetic.

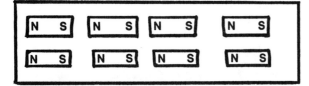
Figure 1. Magnetic

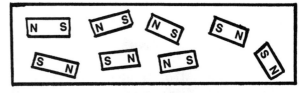
Figure 2. Not magnetic

Soft iron, although magnetic, makes poor magnets. It is not strong and loses its magnetism. However, if iron is combined with carbon and metals such as nickel and chromium, it makes the alloy steel. When magnetized, steel makes excellent magnets.

1. Are all metals magnetic? _____

2. Name the metals that are magnetic. _____

_____

3. These metals are thought to be magnetic because their atoms are aligned and "pointing" in the

   same _____.

4. Soft iron is a weak magnet but when combined with other metals to make the alloy _____

   _____ , it produces a stronger magnet.

5. Is gold magnetic? _____ . Explain your reasoning.

_____

_____.

Applying the information. Place an "N" before substances that are not magnetic and an "M" before substances that are magnetic.

_____ 1. Cobalt   _____ 3. Aluminum foil

_____ 2. Silver   _____ 4. Copper

MATERIALS

Student Worksheet 7-3, "About Magnetism"

INSTRUCTIONAL PROCEDURES

Ask, "According to our last lab, what kinds of things are magnetic?" (Some metals.)

Ask, "How many different metals do you guess are magnetic?" Listen to all answers. Make no judgments.

Explain that scientists can identify the chemical makeup of a substance that is magnetic. Consequently, all of the metals that are magnetic have been identified. Say, "We can learn about these magnetic metals by reading the findings of scientists. Here is a worksheet with information about magnetism."

Student Directions

1. All of the answers to the questions are in the information on the worksheet. Work together with the members of your group to find the answers.
2. Read the questions first. Then read the worksheet, looking for the answers.
3. Write what you think is correct on your own paper.
4. Be prepared to discuss your answers.

DISCUSSION

After 20 to 30 minutes, your students should be ready to discuss the answers to the questions on the worksheet. Answers to Student Worksheet 7–3, "About Magnetism":

1. No
2. Iron, nickel, and cobalt
3. Direction
4. Steel
5. No

Explain your reasoning.

Iron, nickel, and cobalt are the three metals that are magnetic.

Applying the Information: An "N" for nonmagnetic and an "M" for magnetic.

M 1. Cobalt   N 3. Aluminum foil
N 2. Silver   N 4. Copper

Lesson 4. Investigating Magnetism with Paper Clips

OBJECTIVES

Students will answer the questions about the qualities of a magnet by performing tests.

MATERIALS

Student Worksheet 7-4, "Investigating Magnetism with Paper Clips"
On each tray place:
  One magnet
  One box of paper clips

INSTRUCTIONAL PROCEDURES

Distribute the worksheets. Ask someone to read the first question. Then ask for predictions. Continue reading and making predictions for questions one through five. Discuss the directions for six and seven.

Lab Directions

Direct students to use the magnet and the paper clips to answer the questions on the worksheet and to be ready to discuss their discoveries.

DISCUSSION

Discuss the answers to the worksheet.

Answers to Student Worksheet 7-4, "Investigating Magnetism with Paper Clips."

1. Each group will have a different answer depending on the strength of the magnet. Have each group report its answer.
2. Yes.
3. It can hold clips in a chain, bunch, row, and many single ones at a time.
4. Some paper clips become temporary magnets. That is, after a paper clip is removed from the magnet, it is a temporary magnet and capable of magnetically attracting other paper clips.
5. The ends are the stronger parts. The ends are called the poles.
6. Students' observations may vary.
7. Students' questions may vary. Questions can be answered by other students, future labs or lessons, science books, encyclopedias, and the teacher. All questions do not need to be answered during this class. All questions may not be answered. Scientists do not find the answers to all of their questions.

Name _____   Date _____

Investigating Magnetism with Paper Clips

1. How many clips will the magnet hold on one end?

2. Can the magnet hold a "bunch" of paper clips?

3. Can the magnet hold lots of clips in a chain or some other way?

4. What happens to the paper clip after you remove it from the magnet?

5. Is one part of the magnet stronger than the others?

6. List other observations that you made.

7. List any questions that you have based on this investigation.

Lesson 5. Math 'n' Magnets

OBJECTIVES

Students will compute the average number of paper clips a specific magnet can hold.

MATERIALS

Select the table of data that is appropriate for your class from among the three: "The Number of Paper Clips That Our Magnet Holds."

On each tray place:

Two boxes of paper clips

One magnet

INSTRUCTIONAL PROCEDURES

Ask, "Using paper clips, can you devise a way to determine how strong a magnet is?" (See how many clips it will hold.)

Ask students if they think the magnet will hold the same amount of paper clips for each trial. (Probably not.)

Ask students if they remember how to calculate significant numbers and averages. Review the following terms:

1. The mode is the most frequently occurring number.
2. The median is the middle number in a series of numbers arranged in order of value.
3. The mean is the sum of the numbers divided by the quantity of the numbers.
4. The range is the difference between the highest and lowest numbers.

Teaching Tips

1. If your students have forgotten how to identify these significant numbers, invest one class period in review and then move on to the lab.

2. If you did not teach how to statistically analyze data, use the simplest of Student Worksheets, 7-7, and modify the lesson accordingly.

Lab Directions

Hold up one magnet. While you are placing paper clips on one of the poles, say "Put the paper clips on one of the poles. When no more will stick, count the number of clips and record that number on the table of data. Complete five trials and then compute the mode, median, mean, and range and write a conclusion."

Distribute the tables of data and trays of supplies.

After all of the data are gathered and analyzed, have each lab group report on the strength of its magnet.

TEACHING TIME: One or two days. This lab can be repeated to reinforce the math concepts and to reinforce the idea that scientists repeat experiments and tests because replication of results is important.

STUDENT WORKSHEET 7-5

Name _____   Date _____

The Number of Paper Clips That Our Magnet Holds

Trial	No. of Clips	Mode	Median	Mean	Range	Conclusion

Name _____   **Date** _____

The Number of Paper Clips That Our Magnet Holds

Trial	No. of Clips	Mode	Median	Range	Conclusion

Name _____   Date _____

The Number of Paper Clips That Our Magnet Holds

Trial	Number of Paper Clips	Conclusion

Lesson 6. Observing the Lines of Force, Indirectly

OBJECTIVES

Students will:

1. State that magnets have a magnetic force field and nonmagnets do not.
2. Explain that a magnetic force field can be observed by sprinkling iron filings on a magnet.
3. Sketch the line of force for at least one type of magnet.

MATERIALS

For the demonstrations:

An overhead projector

A screen or suitable wall

A pair of bar magnets and a horseshoe magnet

"U" and a circular magnet (Optional)

A metal bar or cylinder that is not a magnet

A large shaker of iron filings

A sheet of plastic, glass, or clear plastic bag

A cup or jar to hold the filings after they have been used

For students:

Paper and pencil for sketching

INSTRUCTIONAL PROCEDURES

Demonstration One

1. Turn on the overhead projector.
2. Cover a magnet with clear glass or place it in a clear plastic bag.
3. Place the bar magnet on the projector.
4. Sprinkle iron filings over the magnet. The filings will line up along the lines of magnetic force of the magnet allowing all to indirectly observe the lines of force.
5. Have someone describe what is on the screen.
6. Introduce new terms: the iron filings line up along the "lines of force" of the magnet revealing the "magnetic field" of the magnet.
7. Ask for predictions when the same procedure is repeated using the metal, nonmagnetic cylinder. Accept all predictions.

Demonstration Two

1. Remove everything from the projector.
2. Pour the used iron filings in a cup.
3. Cover it with plastic or glass.
4. Place the cylinder that is not a magnet on the projector.
5. Sprinkle with unused iron filings. Be sure to use new filings because used ones can become temporary magnets.
6. Have students make observations. (The iron filings stay wherever they land. They are not affected by the cylinder.)
7. Have students make inferences. (The cylinder does not have a magnetic field; therefore, it is not a magnet.)
8. Ask, "How can we test our inference? How can we determine if this is a magnet or not?" (Test it by touching it to items that we know are attracted to magnets. If it attracts them it is a magnet. If it doesn't attract them it isn't a magnet.)
9. Touch the cylinder to paper clips. (It does not attract paper clips. It does not attract the iron filings. It does not have a magnetic field. It is not a magnet.)

Looking at Different Magnetic Fields

1. Put a bar magnet on the projector.
2. Cover it with plastic and sprinkle with iron filings.
3. Direct students to draw what they see. Use a solid line to represent the magnet and dotted lines to represent the lines of force. See Figure 7-1.
4. Repeat the same procedure for a horseshoe magnet. See Figure 7-2.

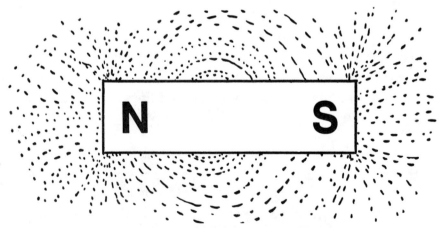

Figure 7-1. Bar magnet and iron filings

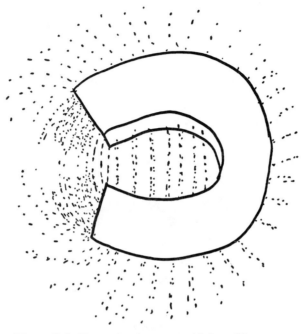

Figure 7-2. Horseshoe magnet with iron filings

Lesson 7. A Second Look at Magnetic Lines of Force

OBJECTIVES

Students will sketch the lines of force observed when the poles of two magnets are brought together and sprinkled with iron filings.

MATERIALS

Overhead projector
Pair of bar magnets
Iron filings
Sheet of plastic
Cup for used iron filings

INSTRUCTIONAL PROCEDURES

1. Place two bar magnets on the projector so that a north and south pole are facing each other. Allow the two magnets to attract each other.
2. Pull them apart.
3. Cover with plastic and sprinkle with iron filings.

4. Direct students to sketch the magnets and the lines of force. See Figure 7-3.

5. Place the magnets so that the north poles are facing each other. Allow the two magnets to push away from each other.

6. Cover with plastic and sprinkle with iron filings.

7. Direct students to sketch the magnets and the lines of force. See Figure 7-4.

8. Place the magnets so that the two south poles are facing each other. Allow them to push apart.

9. Cover with plastic and sprinkle with iron filings.

10. Direct students to sketch the magnet and lines of force. See Figure 7-5.

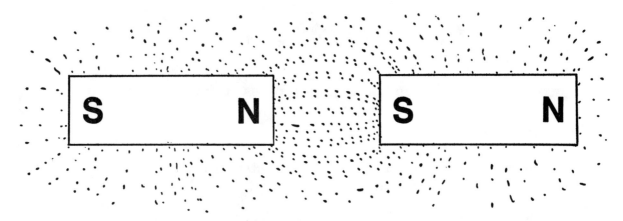

Figure 7-3. Iron filings on N and S poles of two magnets

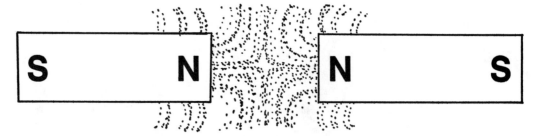

Figure 7-4. Iron filings on N and N poles of two magnets

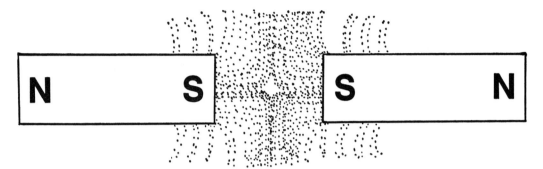

Figure 7-5. Iron filings on S and S poles of two magnets

DISCUSSION

The iron filings indirectly show:

- The lines of force.
- An "attraction" between north and south poles.
- A "repulsion" between like poles.

Lesson 8. The Earth Is Like a Giant Magnet (Grades 4–8)

OBJECTIVES

Students will:

1. Make a compass using a ruler, string, magnet, and stack of books.
2. State that the earth has a magnetic field.
3. Explain that the earth's magnetic field causes compass needles to point in a northerly direction.

MATERIALS

For the demonstration, the same equipment from the last lesson:

 Overhead projector

 Screen or suitable wall

 Bar magnet

 Iron filings

 Plastic sheet

For the lab place the following items on each tray:

  One bar magnet

  One piece of string about eight to ten inches long

  One piece of string about three inches long

  One ruler

INSTRUCTIONAL PROCEDURES

Ask, "Do you know how a freely suspended magnet behaves?" Listen to all inferences. Make no judgments. Explain that they are going to test their inferences and observe freely suspended magnets in today's lab.

Lab Directions

Direct students to use a ruler and string to suspend a magnet from a stack of books. To do so:

- Tie a long piece of string on both ends of a bar magnet.
- Hang the string on a ruler placed in the top of a stack of books. See Figure 7-6.
- Have someone lean on the stack of books to keep the ruler in place.
- Observe the magnet as it twists back and forth and stops.
- Note the alignment, or direction, in which the north pole is pointing. (North)
- Note the direction in which all of the magnets are pointing. (They all point in the same direction, North.)

Figure 7-6. Stack of books to hold a magnet in a swing

Demonstration/Explanation

1. As you turn the projector on, place the bar magnet so that its north pole is pointed to the north.

2. Cover it with plastic and sprinkle with iron filings.

3. Explain that the earth acts as a giant bar magnet. Accordingly, it has the properties of a bar magnet.

4. Ask, "What does that tell you that the earth has in common with a bar magnet?" Acceptable responses:

    a. It attracts other magnets.

    b. It has a magnetic field.

5. Ask, "Why did all of our magnets point in the same direction?" (They were attracted by the Earth's north magnetic pole. They were influenced by the earth's magnetic field. They were all pointing north.)

6. Ask, "What use could be made from this information?" Acceptable answers:

    a. You can always find the north.

    b. You can locate directions.

7. Ask, "Do you know what you made today when you suspended the magnet?" Your students may or may not be able to answer. If they cannot, tell them that they made a compass. The essential part of all compasses is a freely swinging magnet. It is attracted by the earth's north magnetic pole. Early explorers made compasses by suspending natural magnets.

8. Ask, "Do you think that a compass would work on the moon or other planets?" Listen to the comments. Explain that the moon has no magnetic field. As a result no compass would work on the moon. NASA (National Aeronautics and Space Administration) is investigating the possibility of magnetic fields on other planets and other moons.

Teaching Tip

Since any classroom can have some iron pipes in the walls which would attract magnets and not permit them to point north, it is recommended that you use a compass or freely suspended magnet to make sure that it will point north in your classroom. If not, skip this lesson or just explain the concepts.

Lesson 9. Making Another Type of Compass (Optional)

OBJECTIVES

Students will make a compass using a needle, magnet, cork, and container of water.

MATERIALS

On each tray place:
    One plastic container with water in it
    One paper towel
    One cork
    One needle fastened to a piece of cardboard
    One magnet

INSTRUCTIONAL PROCEDURES

1. Review the concept that the earth acts like a giant bar magnet. The north magnetic pole of the earth attracts the north-seeking pole on a magnet. Thus any freely suspended magnet can become a compass and point in a northerly direction.

2. Explain that early explorers like Columbus used magnets to make compasses to help them locate directions. They knew that they could turn certain needles into temporary magnets that would point north by stroking them with lodestone, a natural magnet.

3. Direct students to look at the tray of supplies to determine how they might make an early compass that was used by early sea-going explorers. Emphasize the key ideas are to make a light, temporary magnet by stroking a needle with a magnet and then floating it.

4. Distribute the trays of supplies and observe students as they attempt to make a compass.

DISCUSSION

After the needle is transformed into a temporary magnet, it can rest on a cork which can float on the water. When the movement of the water settles down, the needle will line up in a north-south direction. Columbus and other early explorers made compasses like this.[1]

Lesson 10. Discovering the Law of Magnetism (Grades 4–8)

OBJECTIVES

Students will discover and state the Law of Magnetism: Like poles repel; opposite poles attract.

---

[1] Glenn O. Blough and Julius Schwartz, *Elementary School Science and How to Teach It* (New York: Holt, Rinehart & Winston, 1969), p. 617.

MATERIALS

Student Worksheet 7-8, "Observing Magnetism"
On each tray place:
    One pair of bar magnets with the poles marked N and S
    One string with loops on each end
    An overhead projector

ON THE BOARD

Like poles (N and N or S and S)
Opposite poles (N and S or S and N)
Attract: to draw to itself, to stick
Repel: to push away

INSTRUCTIONAL PROCEDURES

Lab Directions

1. Give out Student Worksheet 7–8, "Observing Magnetism." Go over all of the directions orally.

2. Emphasize that it is necessary to place the magnets flat on the desk to complete directions one through four.

3. Demonstrate how to place a magnet in a "swing" to complete numbers five through eight. See Figure 7-7.

4. Distribute the trays of supplies.

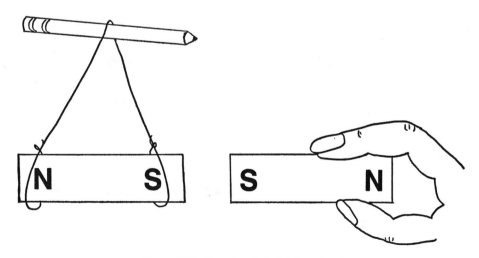

Figure 7-7. A swing to hold a magnet

Name _____   Date _____

Observing Magnetism

Directions: Follow the directions for the placement of magnets; then write your observations in your own words. Next, rewrite your descriptions using the scientific terms **attract**, meaning to pull together and **repel**, meaning to push apart.

Placement of Magnets	Observations in your words	Observations in scientific terms
Place magnets flat on the desk: 1. [S N] [N S] 2. [N S] [S N] 3. [S N] [S N] 4. [N S] [N S] Place one magnet in the string "swing." Bring the pole of the other magnet near it. 5. [S N] [N S] 6. [N S] [S N] 7. [S N] [S N] 8. [N S] [N S]		

In as few words as possible, using the scientific terms listed on the board, write two phrases that describe what happened.

_____

_____

_____

_____

_____

DISCUSSION

Discuss each observation.

Answers to Worksheet 7-8, "Observing Magnetism "

Placement of Magnets	Observations in your words	Observations in scientific terms
1. [S N][N S]	push away	repel
2. [N S][S N]	push away	repel
3. [S N][S N]	come together	attract
4. [N S][N S]	come together	attract
5. [S N][N S]	push away	repel
6. [N S][S N]	push away	repel
7. [S N][S N]	come together	attract
8. [N S][N S]	come together	attract

Two phrases to describe what happened:
Likes repel.
Opposites attract.

Explanation

The phrase "Likes repel; opposites attract," is referred to as the *Law* of Magnetism because all magnets will behave this way. If the N and S markings were filed off, like poles would still repel and opposite would still attract.

Many students will be pleased that they are able to write a phrase that expressed the meaning, if not the exact words, of the Law of Magnetism.

Lesson 11. Making an Electromagnet with a Nonmagnetic Nail

OBJECTIVES

Students will construct an electromagnet using a nonmagnetic nail, battery, and coil of wire.

MATERIALS

Student Worksheet 7-9, "Investigating Electromagnetism"
For the demonstration:

Name _____ **Date** _____

Investigating Electromagnetism

Directions: Make an electromagnet by wrapping a coil of wire around a soft iron nail or cylinder. Connect the two ends of the wire to a 6-volt battery. The iron becomes a temporary magnet when an electric current flows through the coil of wire surrounding the nail.

Table of Data

Is There a Relationship Between the Number of Coils and the Number of Paper Clips That This Electromagnet Holds?

Number of Coils	Number of Paper Clips	Conclusions
3		
6		
9		
12		
15		
18		
21		
24		
27		
30		
33		
36		

One 6-volt battery

Five feet of wire with both ends stripped

A nonmagnetic nail or metal cylinder

Some paper clips

For the lab, place the following items on trays of supplies:

One rectangular 6-volt battery

One piece of bell wire, three to five feet long, with both ends stripped

One iron nail or iron cylinder that is not magnetic

Two boxes of paper clips

INSTRUCTIONAL PROCEDURES

Demonstration

1. Put some paper clips in your hand. Touch them with the nonmagnetic nail. Ask, "Is this nail a magnet?" (No)
2. Make five to ten coils of wire around the nail. Touch the nail to the paper clips. Ask, "Is this a magnet?" (No)
3. Fasten both ends of the wire to the battery. Touch some paper clips with the nail. Ask, "Is this a magnet?" (Yes)
4. Ask if anyone knows what this is called. If no one knows, introduce the term *electromagnet*, a soft iron core that temporarily becomes a magnet when an electric current flows through a coil of wire wrapped around it.
5. Say, "Notice that I made five coils. How could we find out if the number of coils affects the strength of the electromagnet?" (Change the number of coils and count the number of clips that the electromagnet will hold.)

Lab Directions

1. Distribute worksheets. Direct your students to look at the table of data. Ask, "How many coils do you add for each trial?" (3.)

2. Direct students to hold both ends of the wire while working to keep it from flipping about and causing injury.

3. Distribute the trays of supplies.

Discuss Student Worksheet, "Investigating Electromagnetism"

Ask, "Based on the lab work, is there a relationship between the number of coils and the number of paper clips that these electromagnets held?" (Yes. The more coils of wire, the more paper clips it will attract.)

Teaching Tips

If you have taught graphing, consider having students graph the results. For third grade make an electromagnet; eliminate the worksheet.

Lesson 12. Oersted's Serendipitous Discovery (Grades 5–8)

OBJECTIVES

Students will:

1. Discover that a wire carrying an electric current affects the needle of a compass.
2. Explain that Hans Christian Oersted serendipitously, unexpectedly, discovered that electricity flowing through a wire produces a magnetic field.

MATERIALS

On each tray place:

One 6-volt battery

A piece of wire about five feet long with both ends stripped

One compass

Two boxes of paper clips

INSTRUCTIONAL PROCEDURES

Lab Directions

1. Direct students to put the compass on the desk and observe in which direction the needle points.
2. Connect both ends of the wire to the battery, move the wire near the compass and observe the needle. (The needle moves.)
3. Disconnect the battery and observe the compass needle. (The needle returns to its original position.)

Minilecture

While Hans Christian Oersted was connecting a battery for a demonstration in one of his physics classes, he noticed that the wire carrying electricity affected a compass needle. This was an accidental find . . . a serendipitous find.

He knew that the compass needle was a magnet and then suspected that an electric current in the wire produced a magnetic field which affected the magnetic needle. This meant that electricity can produce a magnetic field and an electro-magnet.[2]

Lesson 13. Reading to Identify Electromagnets

OBJECTIVES

Students will list electrical devices that contain electromagnets.

MATERIALS

Texts and encyclopedias that have information about electromagnets and devices containing electromagnets such as electric motors, telephones, telegraphs, and generators.

INSTRUCTIONAL PROCEDURES

Explain that each lab group is a team. Each team uses the science books and encyclopedias to identify as many devices that have electromagnets as possible.

DISCUSSION

After about 30 minutes, ask for one person from each group to read its list. Discuss the workings of the devices identified as containing electromagnets.

An "If You Like" Assignment

Invite students to transfer this information into a poster or a montage showing a variety of uses of electromagnets or actually put together a display of devices containing electromagnets.

Teaching Tips for Teaching a Unit on Magnetism

1. It is recommended that you purchase a High Strength Magnetizer from an educational science supply company. The Magnetizer creates a magnetic field in metals and will revitalize bar and horseshoe magnets. After magnets are dropped a few times, they lose some of their strength and your lessons become less dramatic and may not work. The High Strength Magnetizer will enable you to revitalize your magnets with the flip of a switch.

---

[2] Blough and Schwartz, p. 637.

2. It is also recommended that you purchase a pair of wire cutters from a local hardware store. You need them not only for cutting wire but for stripping insulation off the ends.

3. When ordering batteries, consider how students will attach the wire to the battery. It is easier for students to attach the wires to batteries that have springs or screw terminals. If you must use the cylindrical batteries, without easy-access terminals, provide one strong rubber band with each battery to hold the wires in place.

Most Students Are Attracted to Magnets

Since most students are interested in magnets, you can provide a variety of high-interest lessons. You can present a variety of activities and lessons that teach basic skills in reading, language, math, and science. In doing so, you make the most of your magnets and provide your students with a wealth of information, skills, and successful experiences in problem solving. Your students can become magnet experts, be challenged, and feel good about their accomplishments.

Guided Discovery with Electricity

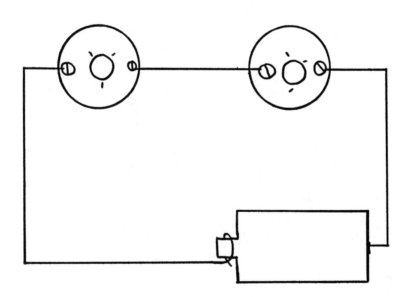

Personal Perspective

Electricity is an integral part of our daily lives. Students use electrical devices daily, but usually do not know how or why they work. Through guided activities with static and current electricity, you can provide your students with a variety of safe, exciting discoveries that will help them better understand their world and build self-confidence. When students successfully wire a light or a bell in a complete circuit, they get immediate feedback . . . it works. Frequently they exclaim, "It works!" or "I did it!" The classroom becomes a place of success for the student and the teacher.

Two ideas about electricity that fascinate my students are:

1. The fact that no one, not even the world's leading scientists, knows exactly what electricity is.

2. No one can see electricity, but we can observe its effects such as the light from a bulb or a picture on a T.V. screen.

Before Teaching about Electricity

Before teaching about electricity it is important for you to understand some basic concepts and safety procedures to make your investigations safe for you and your students.

1. For student investigations use 1½ volt dry cells. They keep the volts and amps (unit for measuring the strength of current) at low, safe levels. Six-volt batteries are also acceptable. Note: Nine volts at three amps can cause burns, shock, and death.

2. A short wire connecting the terminals of a dry cell or battery can cause it to **short out**. That is, it will heat up, can cause a burn, and ruin the cell. Therefore, do not permit students to place a short piece of wire or a paper clip across the terminals.

3. When dry cells or batteries are connected in series, the voltage adds up. Two 1½-volt dry cells connected in series will produce three volts of electricity. See Figure 8-1. It is recommended that when you set up the supply trays, you place one dry cell on each tray. Accordingly, each group receives only one dry cell and will be working with only 1½ volts of electricity.

4. When dry cells are connected in parallel, the voltage does not add up. Thus, two 1½-volt dry cells connected in parallel produce 1½ volts of electricity. See Figure 8-2.

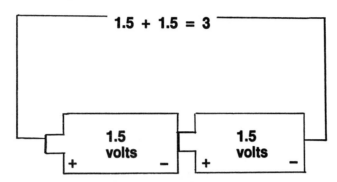

Figure 8-1. Two 1.5 volt dry cells in series produce 3 volts of electricity

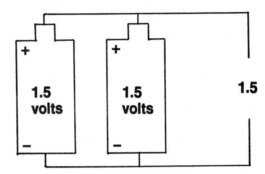

Figure 8-2. Two 1.5 volt dry cells in parallel produce 1.5 volts of electricity

5. A 1½-volt dry cell will light a 1½-volt lamp, and not a 6-volt lamp.

6. Many bells and buzzers require 6 volts. A 6-volt lantern battery is recommended.

7. A 6-volt lantern battery will burn out a 1½-volt lamp in a second or two. If you use a 6-volt lantern battery with lights, use 6-volt lights.

8. All flashlight lamps or small lamps are not necessarily 1½ volts. The required voltage is written on the lamps.

9. A pair of wire cutters is essential for cutting and stripping wire. Using scissors is unsafe and teaches poor habits.

10. When stripping wire, place the wire in the appropriate size hole. Pull the wire up and the cutters down. Since the wire that is being stripped off can fly off and hit someone, it is important to aim away from people. Thus, it is recommended that the teacher cut and strip the wire before class and train students in the safe use of wire cutters.

11. It is recommended that you purchase the type of dry cell that has the two terminals on the top. These are easier for students to wire, do not roll off the desk, and do not need holders.

12. If you already have the "D" and "C" type dry cells that roll and are difficult to wire, use a strong rubber band to keep the wires in place and a piece of masking tape to keep the dry cell from rolling off the desk.

13. If you have never worked with electricity before, you must try the experiments before you teach the lessons.

14. If you are not sure about some of the activities, ask another person who understands electricity for help.

15. If you don't understand or question an activity or its safety, don't teach it until you fully understand it.

16. Static electricity activities require dry weather.

Electricity Is a Flow of Electrons

Even though no one knows exactly what electricity is, scientists know that electricity is a flow of electrons, the negatively charged particles in atoms. Everything is made of atoms. Atoms are made up of electrons, protons, which are positively charged particles, and other particles in the center. The protons do not move easily. Revolving about the center of the atom are electrons, which move easily. If the electrons move from one substance onto another, the substance that loses the electrons has a positive charge, and the substance that receives the electrons has a negative charge. If an atom retains all of its electrons, that atom does not have an electrical charge because it has the same amount of protons and electrons. It is electrically balanced, or neutral.

There are two types of electricity: static and current. Static electricity is the result of a *buildup* or *deficiency* of electrons. Since nature strives for a balance, electrons will jump to the area of deficiency. You can observe this as a spark when you take off a wool sweater or climb under the sheets on a dark dry night. You can observe it in lightning as the charges jump from cloud to cloud, or cloud to ground and back to the cloud, to even out the balance of electrons.

Current electricity is a flow of electrons created when the electrons jump from cloud to ground or move in wires. It is safe for children to investigate current electricity by using dry cells and lantern batteries with low voltage of 6 or less volts and low amps.

Teaching about Electricity

When introducing electricity, stress safety regulations:

1. Never use a wall outlet at home or at school for experiments because they deliver about 110 volts, which can be deadly.

2. When using dry cells or batteries, disconnect the wires if they become warm or hot.

3. Always disconnect dry cells and batteries when not in use.

4. Do not work with more than 6 volts of electricity.

5. Remember, do not touch any science equipment without permission and directions.

Static First? Not Necessarily.

When teaching electricity, teachers often start with static and proceed to current electricity. It is a logical approach, but not essential. It is essential, however, to teach about static electricity during a time of the school year when the weather is dry in order for the static electricity experiments to work. The electrons need dry air in which to build up and make their "jumps." That is why there is more "static cling" when taking clothes out of the dryer on a dry day than on a humid day. However, you can teach about complete circuits on any day because the electrons "jump" through the wire and are not affected by the weather.

Lessons with Static Electricity

Lesson 1. Introducing Static Electricity with Balloons

OBJECTIVES

Students will:

1. State that the two electrical charges are positive and negative.

2. State that a positive and a negative charge attract each other.

3. State that static electricity is a buildup or deficiency of electrons, a negative charge.

MATERIALS

Two balloons blown up and tied off

A comb

A tray containing many small pieces of torn paper

INSTRUCTIONAL PROCEDURES

Demonstration One: A Hair-Raising Experience

1. Select a volunteer who has short, straight hair with no hair spray. Rub the volunteer's hair with a balloon. Then slowly move the balloon near the hair but do not touch the hair. Be prepared for laughs. Ask for observations. (The hair stands on end, is attracted to the balloon.)

2. Repeat with a few more students to replicate the results.

Explanation

The balloon picks up electrons, negative charges, when rubbed on the hair. That gives the balloon a negative charge. The hair lost electrons. That gives the hair a positive charge. Positive and negative charges attract each other.

Demonstration Two: Balloon and Many Small Pieces of Paper

1. Rub the balloon on a volunteer's hair. Move the balloon over the tray containing the pieces of torn paper. Ask for observations. (The paper jumps to the balloon.)

2. Repeat. Move the balloon further from the paper.

3. Tell students that the balloons and paper work in a similar way to the balloon and hair. Ask for explanations.

Explanation

Rubbing the balloon with hair puts extra electrons, or a negative charge, on the balloon. The paper does not have an electrical charge because it has the same number of positive and negative charges on it. It is neutral. The negative charge on the balloon attracts the positive charges in the paper. *Positive and negative charges attract each other.*

Demonstration Three: Comb and Bits of Paper

1. Select a volunteer who has no hair spray. Have that person comb his or her hair and then move the comb over the tray containing bits of paper. Ask for observations. (The paper jumps to the comb.)

2. Ask for explanations. (The negative charge on the comb attracts the positive charges on the paper.)

Putting It Together

- Static electricity is a buildup or deficiency of electrons, a negative charge.
- Positive and negative charges attract each other.

Lesson 2. Another Look at Static Electricity

OBJECTIVES

Students will:

1. State that two negative charges repel each other.

2. Write that like charges repel each other and opposite charges attract.

3. Apply information about static electricity to identify missing charges in static electricity problems.

MATERIALS

Tie an inflated balloon on each end of a string
A piece of wool
A piece of silk
Two glass rods, one tied to a string
A sink with running water and a comb (if possible)
Student Worksheet 8-1, "Static Electricity"

INSTRUCTIONAL PROCEDURES

Demonstration One: Balloons with Wool

1. Have a volunteer hold a string at the midpoint allowing the balloons to hang freely. Ask for observations.

2. Rub each balloon with wool. Move one balloon near the other but do not allow them to touch.

3. Ask for observations. (The balloons push apart.)

4. Ask for inferences. Listen to their ideas.

Explanation

The balloons each pick up electrons or a negative charge from the wool. Negative charges repel each other.

Demonstration Two: Two Glass Rods with Silk

1. Tell the class that you are going to rub the two glass rods with silk and then move one near the other. Ask for predictions.

2. Rub the rods with silk. Hold one rod by the string. Let it swing freely. Move the other close to it but do not allow them to touch. Ask for observations. (They repel each other.)

3. Ask for inferences. (The silk puts the same charge on each rod.)

Explanation

The glass rods each have a positive charge. Two positive charges repel each other. Like charges repel each other.

Demonstration Three: A Glass Rod and a Balloon

1. Rub a glass rod with the silk to give the rod a positive charge.
2. Rub a balloon with wool to give the balloon a negative charge.
3. Explain the charges. Ask for predictions.
4. Hold the balloon by the string. Move the glass rod near it.
5. Ask for observations. The rod and balloon attract each other.
6. Ask for inferences.

Explanation

The wool rubbed extra electrons on the balloon giving the balloon a negative charge. Rubbing the glass rod with silk gave the glass rod a positive charge. Positive and negative charges attract each other.

Demonstration Four: Bending Water

1. Ask students if they think it is possible to bend a stream of water. (Most will think that it is not possible.)
2. Comb a volunteer's hair.
3. Turn the classroom faucet on to produce a slow but steady stream.
4. Move the comb up and down near the stream of water.
5. Ask for observations. (The water bends!)
6. Ask for inferences. If someone can explain what is happening, great. If not, explain.

Explanation

Water does not have a charge. It has a balance of positive and negative charges. The comb picked up extra electrons, negative charges, from the hair. The negative charges on the comb repel the negative charges in the water and attract the positive charges in the water.

"If You Like" Assignment

Try "bending" water at home to "shock" and amaze your family.

PRACTICE AND REVIEW

Distribute Student Worksheet 8-1, "Static Electricity." Direct students to fill in the missing charges on the drawings.

After five or ten minutes, review the answers with students. Direct them to change any charge that needs to be changed and make sure that they write that "Like charges repel and opposite charges attract."

Name _____   **Date** _____

Static Electricity

Directions: Fill in the missing charge.

1.

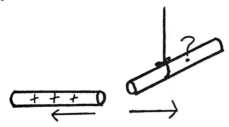

2.

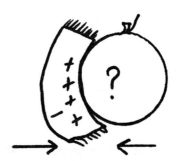

3.

4.

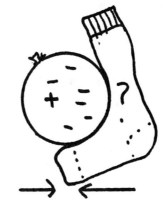

5.

6.

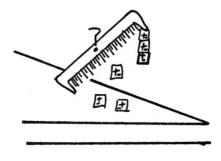

Directions: Fill in the blanks using the words "attract" and "repel."

7. Like charges _____

8. Opposite charges _____

Answers to Student Worksheet 8-1, "Static Electricity"

1. Positive	4. Positive	7. Repel
2. Negative	5. Positive	8. Attract (The paper has no charge. It has a balance of positive and negative charges. The negative charges on the comb attract the positive charges on the paper.)
3. Negative	6. Negative	

Lessons with Current Electricity

Lesson 3. Discovering Open and Closed Circuits with Dry Cells and Lights

OBJECTIVES

Students will:

1. Discover that an open circuit does not provide a complete path for electricity and a light will not glow.
2. Discover that a closed circuit provides a complete path for electricity to make a light glow.

MATERIALS

On each tray:
   One "D" size battery (1.5 volts)
   One light (1.5 volts)
   One piece of bell wire, about six inches long, with both ends stripped

INSTRUCTIONAL PROCEDURES

Lab Directions

Direct students to:

1. Use the dry cell and wire and find at least two ways to light the bulb.
2. Sketch the circuits that make the light glow.
3. Signal by raising a hand that the light glows and the sketch is completed.

TEACHER DIRECTIONS

When a group signals that it is finished:

1. Have the group demonstrate two ways to make the light glow.

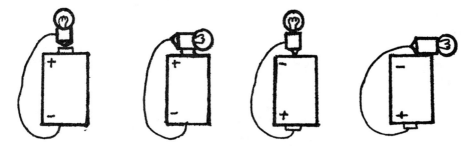

Figure 8-3. Four basic ways to wire a bulb with one wire

2. Encourage them to discover two additional ways to make the light glow.
3. Introduce the terms "closed" and "open" circuits.
   - A closed circuit provides a complete path for the electricity and the light glows.
   - An open circuit does not provide a complete path for the electricity and the light does not glow.

Lesson 4. Opening and Closing a Circuit to Send Morse Code

OBJECTIVES

Students will send at least two short messages in Morse Code using an open and a closed circuit.

MATERIALS

Student Information Sheet 8-A, "The International Morse Code"
On each tray:
   Dry cell
   Light
   One wire

ON THE BOARD

```
S  |  O   |  S
•••|  ___ | •••
```

The International Morse Code distress signal.

Name _____   **Date** _____

The International Morse Code

A	. _	N	_ .	1	. _ _ _ _
B	_ . . .	O	_ _ _	2	. . _ _ _
C	_ . _ .	P	. _ _ .	3	. . . _ _
D	_ . .	Q	_ _ . _	4	. . . . _
E	.	R	. _ .	5	. . . . .
F	. . _ .	S	. . .	6	_ . . . .
G	_ _ .	T	_	7	_ _ . . .
H	. . . .	U	. . _	8	_ _ _ . .
I	. .	V	. . . _	9	_ _ _ _ .
J	. _ _ _	W	. _ _	0	_ _ _ _ _
K	_ . _	X	_ . . _		
L	. _ . .	Y	_ . _ _		
M	_ _	Z	_ _ . .		

INSTRUCTIONAL PROCEDURES

Introduce the Morse Code. Explain that by using open and closed circuits, it is possible to send messages in Morse Code. Demonstrate a brief message, "Hi."

Lab Directions

Direct students to:

1. Set up a circuit using one wire, one light, and a dry cell.
2. Open and close the circuit to make dots and dashes needed to send an S O S signal.
3. Write and send at least one short message using the Morse Code.
4. Send as many messages as time permits.

"If You Like" Assignments

1. Write a report about Samuel F. B. Morse.
2. Make a bulletin board about Morse and the telegraph.
3. Build a telegraph and prepare a message.

Lesson 5: Open and Close Circuits Using a Light in a Receptacle and a Bell or a Buzzer (Optional for Reinforcement)

OBJECTIVES

Students will:

1. Discover how to wire a lamp in a receptacle as a part of a complete circuit.
2. Wire a bell or a buzzer in a complete circuit.

MATERIALS

On each tray:
    One 6-volt battery
    Two pieces of bell wire
    One 6-volt lamp
    One lamp receptacle

For later in the lab:

One bell or buzzer per group (Note: these usually require 6 volts.)

INSTRUCTIONAL PROCEDURES

Lab Directions

Direct students to:

1. Use all of the equipment on the tray to make the light work.
2. Trade the light for a bell or buzzer and make it work.

DISCUSSION

Electricity needs a complete path to and from the power source to make a light, bell, or buzzer work. Thus, electrical devices and appliances perform when placed in a complete or closed circuit. They do not perform when the circuit is opened.

Lesson 6. Discovering How to Use a Knife Switch in a Complete Circuit

OBJECTIVES

Students will discover how to open and close a circuit using a knife switch.

MATERIALS

On each tray:
  One 6-volt lantern battery
  One 6-volt lamp
  One receptacle
  One knife switch
  Three wires

INSTRUCTIONAL PROCEDURES

Lab Directions

Direct students to:

1. Use all of the equipment on the tray to make the lamp glow.
2. Open and close (or turn the light off and on) by using the knife switch.
3. Sketch the circuit.

Lesson 7. Identifying and Classifying
Conductors and Nonconductors

OBJECTIVES

Students will:

1. Build a three-wire electrical tester.
2. Classify a group of items as conductors and nonconductors of electricity based on whether they complete a circuit or not.

MATERIALS

Student Worksheet 8-2, "Classifying Electrical Conductors and Nonconductors"

On each tray the following items to build a three-wire electrical tester:

One battery

Three pieces of bell wire

One lamp

One receptacle

Also on each tray place a variety of items to test.

Suggested items:

Conductors	Nonconductors
Aluminum foil	Wooden popsicle stick
Paper clip	Plastic cube
Metal screw	Rubber band

BEFORE CLASS

Use a dry cell, a lamp in a receptacle, and three wires to make an electrical tester. See Figure 8-4.

INSTRUCTIONAL PROCEDURES

Lab Directions

Direct students to:

1. Use the electrical tester:

    a. touch an item with both wires of the tester

Name _____   Date _____

Classifying Electrical Conductors and Nonconductors

Conductors	Nonconductors

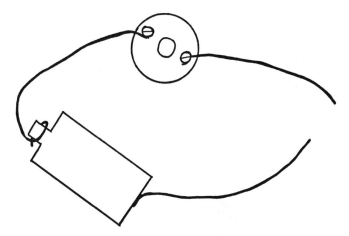

Figure 8-4. A three-wire electrical tester

    b. if the light glows, the item is a *conductor*

    c. if the light does not glow the item is a *nonconductor*.

2. Record the results on the worksheet.

Teaching Tips

Touch the wires of the tester to make sure it works.
Repeat this lab using different items to test.

Lesson 8. What Types of Materials Conduct Electricity?

OBJECTIVES

Students will:

1. Discover that metals and graphite conduct electricity.
2. Discover that nonmetals do not conduct electricity.

MATERIALS

Student Worksheet 8-3, "What Types of Materials Conduct Electricity?"
On each tray place:
    Materials to make an electrical tester. See Lesson 7.
    Suggested materials to test are listed on page 275.

Name _____ Date _____

What Types of Materials Conduct Electricity?

Conductors	Nonconductors	Conclusions

Conductors	Nonconductors
A copper penny	A rubber band
A stainless steel paper clip	A piece of paper
An aluminum bolt	A piece of limestone chalk
The metal part of a wire	The plastic covering of a wire
Graphite in a pencil	Wood in a pencil

ON THE BOARD

List the descriptive names of the items such as aluminum bolt.

INSTRUCTIONAL PROCEDURES

Lab Directions

Direct students to:

1. Build an electrical tester.
2. Test each item to determine if it is a conductor or nonconductor.
3. Record the name and composition of the item on the worksheet.
4. Write a statement that describes the types of materials that conduct electricity.

Acceptable Conclusions

1. Metals conduct electricity.
2. Graphite conducts electricity.
3. Nonmetals do not conduct electricity. (Graphite is an exception.)

Introduce Terms After Lab Experience

Nonconductors are called *insulators*.

Lesson 9. Discovering the Difference Between Series and Parallel (Grades 4–8)

OBJECTIVES

Students will:

1. Complete a circuit using three wires, a dry cell, and two lamps.
2. Complete a circuit using four wires, a dry cell, and two lamps.
3. Describe at least two differences in the two types of wiring.

MATERIALS

On each tray:

One 6-volt lantern battery

Two 6-volt lamps

Two lamp receptacles

Three pieces of bell wire

For later in the lab:

A fourth wire

INSTRUCTIONAL PROCEDURES

Lab Directions

First set of student directions:

1. Use all of the equipment on the tray to make both of the lights glow.
2. Diagram the circuit.
3. Raise hands to signal that the group is ready to:
    a. Demonstrate the completed circuit.
    b. Show the diagrams.
    c. Listen to the next set of directions.

Second set of directions:

1. Make both lights glow using a fourth wire and a different pattern of wiring that is not a larger circle and has only one wire touching each end of the dry cell.
2. Diagram the circuit.
3. Raise hands to signal that the group is ready to:
    a. Demonstrate the wiring.
    b. Show the diagram.

Introduce the Terms Series and Parallel

Diagram the circuit showing three wires. Point out that the wires and lights are wired one after another in a series. This type of circuit is called a *series* circuit; or is said to be wired in series. See Figure 8-5.

Diagram the circuit showing the four wires. Point out that the wires are drawn parallel to each other. That is the lines are equal distance from each other like railroad tracks. This type of a circuit is called a *parallel* circuit. See Figure 8-6.

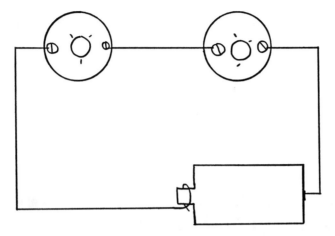

Figure 8-5. Two lights wired in series

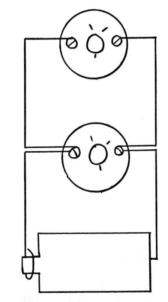

Figure 8-6. Two lights wired in parallel

Teaching Tip

When teaching abstract ideas, it is important to present a concrete example before introducing the terms. When vocabulary is first introduced, students have no previous experience on which to build.

Teaching Time

This lesson usually requires two days.

Lesson 10. Investigating Series and Parallel Circuits (Grades 4–8)

OBJECTIVES

Students will:

1. Identify observable differences in lights wired in series and in parallel.
2. Infer the wiring of a string of Christmas tree lights and give a description of how they work.

MATERIALS

Student Worksheet 8-4, "Investigating Series and Parallel Circuits"
On each tray:
   One 6-volt lantern battery
   Two 6-volt lamps
   Two receptacles
   Four pieces of wire

ON THE BOARD

Sketch Figures 8-5 and 8-6.

INSTRUCTIONAL PROCEDURES

Lab Directions

Direct students to:

1. Follow the diagrams on the board to wire the lights in series.
2. Answer questions one and two.
3. Follow the diagrams on the board to wire the lights in parallel.
4. Answer questions three and four.
5. Use what you know about series and parallel circuits to answer questions five and six.

Answers to Worksheet 8-4, "Investigating Series and Parallel Circuits"

1. Both lights go out.
2. Both lights go out.
3. The first light goes out; the second stays lit.
4. The first light stays lit; the second goes out.

Student Worksheet 8-4

Name _____ Date _____

Investigating Series and Parallel Circuits

Directions: Use three wires to wire two lights in series.
Observe the circuit and answer the following questions.

1. In series, what happens when the first light is unfastened?

   _____

   _____

2. What happens when the second light is unfastened?

   _____

   _____

Directions: Use four wires to wire the two lights in parallel.
Observe the circuit and answer the following questions.

3. In parallel, what happens when the first light is unfastened?

   _____

   _____

4. What happens when the second light is unfastened?

   _____

   _____

5. In some Christmas tree lights, when one light burns out, the entire string of lights goes out.

   How do you think these lights are wired? _____

   Give your reasoning. _____

   _____

   _____

6. In other Christmas tree lights, when one light burns out, only that light goes out and the rest

   of the lights remain lit. How do you think these lights are wired? _____

   Give your reasoning. _____

   _____

   _____

5. In series.

Reasoning: When our two lights were wired in series, and one light was unfastened, both lights went out.

6. Parallel.

Reasoning: When our two lights were wired in parallel, and one light was unfastened, the other stayed lit.

Teaching Time

This lesson may require two days.

Lesson 11. Demonstrate How to Make a Question-and-Answer Box (Grades 4–8)

OBJECTIVES

Students will explain how to make a Question-and-Answer Box.

MATERIALS

A shoe box

One battery (1½- or 6-volt system will work)

One lamp in a receptacle

Five wires

Four brass head fasteners

Four blank labels

Before class

Make a Question-and-Answer Box. See Figure 8-7.

INSTRUCTIONAL PROCEDURES

1. Show the sample Question-and-Answer Box.

2. Touch one wire to the brass head fastener at question one and the other wire to the brass head fastener near the correct answer.

3. Ask for observations. (The light goes on.)

4. Touch the second wire to the "wrong answer."

5. Ask for observations. (The light does not go on.)

6. Open the box; show the wiring for the tester.

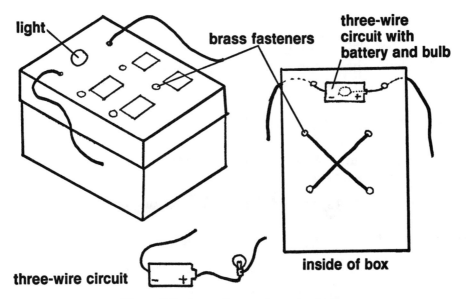

Figure 8-7. A question-and-answer box

7. Ask students how the lid must be wired based on their observations. (A wire goes from the question to its answer.)

8. Show the wiring of the lid.

9. Discuss a variety of other materials that could be used:

    a. peg board in place of the box

    b. nuts and bolts in place of brass head fasteners

    c. a bell or buzzer in place of the light

Note: Some brass head fasteners are coated with plastic, making them nonconductors. You need to make students aware of this.

Lesson 12. Students Make a Question-and-Answer Box

OBJECTIVES

Students will make a question-and-answer box.

MATERIALS

On each tray:
    One battery
    One lamp
    One receptacle

Six pieces of wire
Six brass head fasteners
Six blank labels (Optional)
One box (For younger students, punch holes in the lid.)

INSTRUCTIONAL PROCEDURES

Review how to make a Question-and-Answer Box by asking questions.
Direct students to:

1. Complete a Question-and-Answer Box, making up three questions on any subject.
2. Trade the completed box with a group when finished.

"If You Like" Assignment

Build your own Question-and-Answer Box at home and bring it to class.

Lesson 13. Making a Simple Galvanometer (Grades 6–8)

OBJECTIVES

Students will construct a simple galvanometer using a coil of wire and a directional compass.

MATERIALS

Student Worksheet 8-5, "Making a Galvanometer"
On each tray:
    Two pieces of bell wire, each about three feet long
    A compass
    6-volt battery

INSTRUCTIONAL PROCEDURES

1. Distribute Student Worksheet 8-5, "Making a Galvanometer."
2. Direct students to raise their hands when they can define a galvanometer according to the worksheet. (A meter to detect the presence of small currents of electricity.)
3. Stress that it is essential to control the wire while wrapping, for safety reasons.

Name _____   **Date** _____

Making a Galvanometer

A galvanometer is a meter used to detect the presence of small currents of electricity. A simple galvanometer can be made from several feet of wire and a magnetic compass.

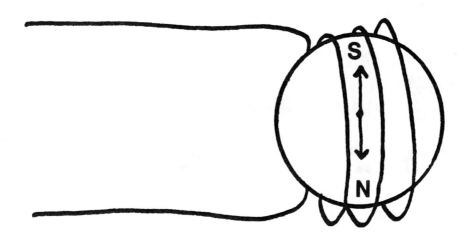

<u>You will need:</u>

- Two strands of wire, each about three feet in length.
- A directional compass.
- A battery.

<u>To make a simple galvanometer:</u>

A. Have a lab partner hold one end of the wire to prevent it from flopping about causing a safety problem.
B. Wrap the wire around the compass, leaving both ends free.

<u>To test the galvanometer:</u>

1. Line up the compass so that the magnetic needle is parallel to the coils of wire wrapped around it.
2. Touch both ends of wire to the terminals on the battery.

3. What do you observe? _____

   _____

4. Touch the ends of the wire to the opposite terminals.

5. What do you observe? _____

   _____

   _____

4. Ask how to make a galvanometer according to the worksheet. (Wrap wire around a compass.)

5. Ask how to test a galvanometer? (Connect it to a dry cell and observe the behavior of a nearby compass.)

6. Direct students to test their galvanometer and record the results on the worksheet.

Answers for Worksheet 8-5, "Making a Galvanometer"

3. The compass needle moves.

5. The compass needle moves in the opposite direction.

Lesson 14. Using a Galvanometer to Investigate the Effects of a Magnet on a Coil of Wire (Grades 6–8)

OBJECTIVES

Students will:

1. Use a galvanometer to determine if a magnet can produce electricity.

2. Explain that Michael Faraday discovered that a magnet can produce electricity.

3. Explain that modern generators produce electricity by moving a coil of wire near a magnet.

MATERIALS

Student Worksheet 8-6, "Using a Galvanometer to Investigate the Effects of a Magnet on a Coil of Wire"

On each tray:

A bar magnet

Two sections of wire

A compass

An empty cardboard cylinder from paper towels or toilet tissue

INSTRUCTIONAL PROCEDURES

1. Distribute Student Worksheet 8–6, "Using a Galvanometer to Investigate the Effects of a Magnet on a Coil of Wire."

2. Review the directions with the class.

3. Distribute the supplies.

4. Discuss the observations.

Name _____   **Date** _____

Using a Galvanometer to Investigate
the Effects of a Magnet on a Coil of Wire

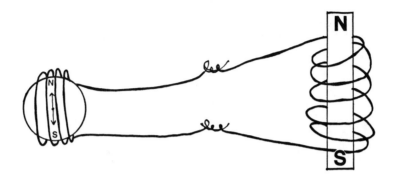

To set up the experiment:

A. Wrap one section of wire tightly around the compass to make a simple galvanometer.

B. Wrap the other section of wire loosely around the cardboard cylinder to make a coil of wire and remove the cylinder.

C. Connect the ends of the galvanometer to the ends of the coil of wire.

Directions:
As you follow each set of directions observe the galvanometer and write your observations.

1. Place the magnet near the coil of wire. Move it back and forth. What do you observe?

_____

2. Place the magnet in the coil. Move it back and forth. What do you observe?

_____

3. Keep the magnet in the coil. Move the coil back and forth. What do you observe?

_____

4. Place the coil over the magnet. Move the coil. What do you observe?

_____

5. Based on your observations, what inferences can you make about the effects of a magnet on a

coil of wire? _____

_____

_____

*Answers to Worksheet 8-6, "Using a Galvanometer to Investigate
the Effects of a Magnet on a Coil of Wire"*

1. The needle moves back and forth.

2. The needle moves back and forth.

3. The needle moves back and forth.

4. The needle moves back and forth.

5. The movement of a *magnet* near a coil of wire produces electricity which is detected by the galvanometer and the movement of a *coil* of wire near a magnet produces electricity.

Minilecture

In 1831, a scientist named Michael Faraday discovered that magnetism can produce electricity by moving a coil of wire near a magnet. His discovery later led to the development of the electric generator in which steam power is used to turn turbines, which move giant magnets inside huge coils of wire.

Most of our electricity comes from power plants which burn fossil fuel, which makes energy that turns a magnet or a coil. In nuclear power plants, nuclear fuel is used to produce heat, which produces steam, which turns a magnet or a coil to produce electricity. Faraday's discovery has been put to everyday use.

Potential Resource

Many power companies have an educational director and/or packets of information and electrical equipment for schools. They are particularly effective in explaining:

1. How electricity is generated.

2. Safety procedures in the home.

3. Safety guidelines for children such as:
    *   staying away from downed power lines
    *   having the power company retrieve cats and kites from power poles and lines.

4. How to compute the cost of operating electrical appliances.

Be aware that those power companies which operate nuclear power plants may give a pro-nuclear power message. That is acceptable. But, it is recommended that the teacher present a balanced view by explaining some of the problems. Consider some of the concerns caused by the incidents at Three Mile Island and Chernobyl:

1. Radiation leaks into the air.

2. Returning hot water to local streams and rivers.

3. Possible meltdowns.

4. Possible long-term damage to the environment, especially underground water supplies.

5. Possible health threats to people.

Teaching Considerations

If you are new to teaching activity lessons about electricity, it is recommended that you consider three groups of lessons presented in this chapter:

1. Static electricity: Lessons one and two.

2. Open and closed circuits: Lessons three and four.

3. Conductors and nonconductors: Lesson seven.

As you build your experience and equipment, you can add other lessons from this section and make up ones of your own.

Beyond the Classroom

Consider inviting those who have erector sets to make electrical devices, bring them to class, demonstrate, and explain how to make them. One of my third grade students made a red robotic cat and a variety of futuristic cars which brightened our classroom and gave him an opportunity to excel. His success further motivated him and he motivated other students.

It is rewarding to build things, especially things that light up, make noise, and make something work. It is even more rewarding to teach someone to discover how to build and how to discover possible solutions when wires get crossed or lights burn out.

I hope that you will experience the joys of science discovery teaching as I have done.

Additional Science Teaching Resources

SCIENCE PROJECTS

by Phil Schlemmer

For grades 4-8, here are 26 detailed, ready-to-use Student Research Projects that teach basic information about a variety of science topics and help develop independent learning skills in seven major areas: research, writing, planning, problem-solving, self-discipline, self-evaluation, and presentation. Each project includes step-by-step Teacher Directions and reproducible Student Handouts you can copy as many times as needed for individual or group use. Among the topics covered are Invertebrate Animals, Animal Taxonomy, Chemistry, Energy and Electricity, Pulleys, Oceanography and Entomology.

Item #C-5094-2
Price $19.95

ELEMENTARY SCIENCE ACTIVITIES FOR EVERY MONTH OF THE SCHOOL YEAR

by Dorothea Allen

Here are scores of stimulating science projects and activities from September through June, with each monthly chapter organized around an important central theme. For example, September features developing "science awareness" by working with jumping beans, bog foliage and the stars . . . February focuses on learning to test hypotheses by determining the effect of salt water on living tissues and the direction of plant root growth . . . and April activities explore practical applications of science in desalting water and making cheese.

Item #259952
Price $21.95

SCIENCE LEARNING CENTERS FOR THE PRIMARY GRADES

by Carol A. Poppe and Nancy A. Van Matre

This resource provides detailed directions to help the teacher develop and organize a system in which each child participates daily in a science learning center . . . plus 40 stimulating learning center activities ready for instant use. Included are five complete science learning center units that will fit into any primary curriculum: THE FIVE SENSES ("Touch," "Tasty Ice Cream Cone") . . . THE HUMAN BODY ("Heart," "Measure Me Neat") . . . SPACE ("The Solar System," "Gravity") . . . PLANTS ("Parts of a Plant") . . . and DINOSAURS ("What Happened to Dinosaur Bones?").

Item #C-7496-7
Price $21.95

READY-TO-USE SCIENCE ACTIVITIES FOR THE ELEMENTARY GRADES

by Debra L. Seabury and Susan L. Peeples

For classroom teachers and science specialists in grades 3-8, here are teacher-directed activities and aids and over 130 reproducible student activities to spark interest and involvement in six major areas of life and earth science: Plants, Animals, The Human Body, Geology, Weather, and Space. Sample activity titles include "Seed Distribution," "A Fish's Body," "Parts of the Eye," "The Water Cycle," and "All About Shooting Stars."

Item #C-7437-1
Price $21.95

SCIENCE DEMONSTRATIONS FOR THE ELEMENTARY CLASSROOM

by Dorothea Allen

This unique resource presents over 100 exciting demonstration lessons that sharpen students' observation and analysis skills and provide a fascinating hands-on introduction to Physical, Environmental and Life Science. These demonstrations give life to dozens of important concepts such as buoyancy, convection and oxidation yet are easy to do and require only inexpensive, readily available materials. Each includes a descriptive title, learning objectives, list of needed materials, step-by-step directions for carrying it out, and a clear scientific explanation of how and why it works.

Item #794651
Price $21.95

SCIENCE DISCOVERY ACTIVITIES KIT: Ready-to-Use Lessons and Worksheets for Grades 3-8

by Frances Bartlett Barhydt

This remakable resource provides over 100 ready-to-use guided discovery science lessons in five major areas: Chemical & Physical Changes, Graphing, Magnets, The Human Body, and Electricity. The lessons teach students how to work and think like scientists and include complete teacher directions along with reproducible student activity and information sheets. Sample lessons include "Why Do Leaves Turn Color in the Fall?", "What Does the Body Do with Digested Food?", "The Effects of Tobacco Smoke on the Respiratory System," and "What Types of Metals Are Magnetic?"

Item #C-7852-1
Price $24.95

K-3 SCIENCE ACTIVITIES KIT

by Carol A. Poppe and Nancy A. Van Matre

In this "kit," the authors of Science Learning Centers offer five more science enrichment units to help primary teachers extend the curriculum and involve students in exploring science independently. Each unit focuses on a single theme and contains eight ready-to-use activities that reinforce skills in science as well as other subject areas such as writing, art, math and reading. The units and sample activities include WEATHER ("Lightning") . . . NUTRITION ("Four Food Groups") . . . BIRDS ("Birds Have Beaks") . . . TREES ("Parts of Trees") . . . and PETS ("Animal Body Coverings").

Item #C4772-4
Price $21.95

LIFE SCIENCE LABS KIT

by Michael F. Fleming

For general-science and biology teachers in grades 7-12, here are 20 ready-to-use, self-contained "labs" for teaching basic concepts in three exciting areas of life science today—Human Anatomy, Genetics, and Health. Each lab focuses on a single topic such as "Mitosis" and "Body Mechanisms Against Infection" and is printed in a handy spiral-bound format for easy copying of full-page models and worksheets. All labs provide specific learning objectives, teaching guidelines, a reproducible model, investigation activities, follow-up questions (and answers), and a self-evaluation sheet to allow students to measure their own learning.

Item #C-540X-6
Price $24.95

SCIENCE & MATH ENRICHMENT ACTIVITIES FOR THE PRIMARY GRADES

by Elizabeth Crosby Stull and Carol Lewis Price

Here's a treasury of stimulating individual and group activities to help children relate science and math to other curriculum areas and develop their awareness of both subjects in the everyday environment. Included are over 350 activities in 16 topical sections, each providing 10 or more reproducible activity pages that can be photocopied as is for instant use. Sample sections include Birds, Fruits & Vegetables, Insects, Monsters, Rodents, Sea Life, Tools, Transportation, Wind, Computers & Robots.

Item #C-7461-1
Price $21.95

SCIENCE CURRICULUM ACTIVITIES LIBRARY

by Marvin N. Tolman and James O. Morton

Organized for easy use in three separately printed, self-contained volumes, this unique "library" provides nearly 500 exciting hands-on activities to teach students thinking and reasoning skills along with basic science concepts and facts. Each volume features one major area of science, presents a store of activities that can be used with any elementary science text, and begins with a variety of Starter Ideas to introduce students to the discovery/inquiry approach.

Book I   LIFE SCIENCE ACTIVITIES FOR GRADES 2-8

Over 180 hands-on experiences covering the topics of Plants & Seeds, Animals, Animal Life Cycles, Animal Adaptation, Body Structure, The Five Senses, Health & Nutrition.

Item #536060
Price $21.95

Book II   PHYSICAL SCIENCE ACTIVITIES FOR GRADES 2-8

Over 170 easy-to-use discovery activities featuring The Nature of Matter, Energy, Light, Sound, Simple Machines, Magnetism, Static Electricity, and Current Electricity.

Item #669796
Price $21.95

Book III   EARTH SCIENCE ACTIVITIES FOR GRADES 2-8

Over 160 hands-on activities covering the topics of Air, Water, Weather, The Earth, Ecology, Above the Earth, and Beyond the Earth.

Item #222521
Price $21.95

NOTE: If order forms are missing, please send your name, address, and the title, item number and price of each of the books you want to:

Prentice Hall
Book Distribution Center
Route 59 at Brookhill Drive
West Nyack, NY 10995-9901